Periodic Precipitation

Related Pergamon Titles of Interest

Books

IUPAC
Solubility Data Series

MASUDA & TAKAHASHI
Aerosols: Science, Industry, Health and Environment (2 vols.)

TAYA
Metal Matrix Composites: Thermomechanical Behaviour

Journals

Journal of Aerosol Science

Materials Research Bulletin

Progress in Crystal Growth and Characterization

Progress in Solid State Chemistry

Solid State Communications

Solid State Electronics

Full details of all Pergamon publications/free specimen copy of any Pergamon journal available on request from your nearest Pergamon office.

Periodic Precipitation

A Microcomputer Analysis of Transport and Reaction Processes in Diffusion Media, with Software Development

HEINZ K. HENISCH
The Pennsylvania State University
University Park, Pa.

PERGAMON PRESS
Member of Maxwell Macmillan Pergamon Publishing Corporation
OXFORD · NEW YORK · BEIJING · FRANKFURT
SÃO PAULO · SYDNEY · TOKYO · TORONTO

U.K.	Pergamon Press plc, Headington Hill Hall, Oxford OX3 0BW, England
U.S.A.	Pergamon Press, Inc., Maxwell House, Fairview Park, Elmsford, New York 10523, U.S.A.
PEOPLE'S REPUBLIC OF CHINA	Pergamon Press, Room 4037, Qianmen Hotel, Beijing, People's Republic of China
FEDERAL REPUBLIC OF GERMANY	Pergamon Press GmbH, Hammerweg 6, D-6242 Kronberg, Federal Republic of Germany
BRAZIL	Pergamon Editora Ltda, Rua Eça de Queiros, 346, CEP 04011, Paraiso, São Paulo, Brazil
AUSTRALIA	Pergamon Press Australia Pty Ltd., P.O. Box 544, Potts Point, N.S.W. 2011, Australia
JAPAN	Pergamon Press, 5th Floor, Matsuoka Central Building, 1-7-1 Nishishinjuku, Shinjuku-ku, Tokyo 160, Japan
CANADA	Pergamon Press Canada Ltd., Suite No. 271, 253 College Street, Toronto, Ontario, Canada M5T 1R5

Copyright © 1991 Pergamon Press plc

All Rights Reserved. No part of this publication may be reproduced, stored in a retrieval system or transmitted in any form or by any means: electronic, electrostatic, magnetic tape, mechanical, photocopying, recording or otherwise, without permission in writing from the publisher.

First edition 1991

Library of Congress Cataloging-in-Publication Data
Henisch, Heinz K.
Periodic precipitation: a microcomputer analysis of transport and reaction processes in diffusion media, with software development / Heinz K. Henisch.
p. cm.
1. Precipitation (Chemistry)--Data processing.
2. Precipitation (Chemistry)--Mathematical models.
I. Title.
QD547.H46 1990 541.3'485--dc200 90-43057

British Library Cataloguing in Publication Data
Henisch, Heinz K. (Heinz Kurt)
Periodic precipitation: a microcomputer analysis of transport and reaction processes in diffusion media, with software development.
1. Chemical engineering. Diffusion
I. Title
660.28423
ISBN 0-08-040276-3

Printed in Great Britain by BPCC Wheatons Ltd, Exeter

To Bridget

LIST OF CONTENTS

1. INTRODUCTION

1.1 General Considerations.	1
1.2 Fick's Laws of Diffusion.	7
1.3 Modified Diffusion Laws.	8
1.4 Basic Diffusion Algorithms.	11

2. CONCENTRATION PROFILES

2.1 Test of the Simplest Diffusion Algorithm.	17
2.2 Diffusion in Non-homogeneous Media.	20
2.3 Diffusion in the Presence of Electric Fields.	24
2.4 Double Diffusion; Concentration Products.	25

3. NUCLEATION ALGORITHMS

3.1 Solubility Relationships in Binary Systems.	29
3.2 Stoichiometric Considerations.	31
3.3 Graincount and Grain Size; Concentration Decrements.	33

4. PRECIPITATE GROWTH AND RE-SOLUTION

4.1 Precipitate - Solution Interaction; Rates of Mass Transfer.	37
4.2 Development of Periodic Structures.	41
4.3 Spacing Relationships.	44
4.4 Deterministic and Probabilistic Models.	46
4.5 Effect of Secondary Reaction Products.	51
4.6 Reservoir Depletion, Contamination and Closure.	53

5. HYPOTHETICAL EXPERIMENTATION WITH BINARY SYSTEMS

5.1 Typical Result Configurations.	58
5.2 Experiments with System Structures.	61
5.3 Experiments with Reservoir Concentration Parameters..	62
5.4 Experiments with Diffusion Coefficients.	67
5.5 Experiments with Solubilities and Precipitation Thresholds.	68

5.6 Experiments with a Secondary Reaction Product.	70
5.7 Experiments with the Stoichiometry Conditions.	73
5.8 Experiments with Electric Fields.	76
5.9 Experiments in the Probabilistic Mode.	79

6. PERIODIC PRECIPITATION IN MONOMER SYSTEMS

6.1 Monomer Precipitation by Solubility Modulation.	83
6.2 Banding of Particle Distributions by Ostwald Ripening.	86
6.3 Experiments with Competitive Particle Growth; a Typical Run.	90
6.4 Experiments with Competitive Particle Growth; Effect of Parameter Variations.	94

APPENDIX A: Precipitation by Binary Reaction; Practical Software Implementation (Programs PPBIN**).	103
APPENDIX B: Monomer Precipitation by Solubility Modulation; Practical Software Implementation (Programs PPMON**).	109
APPENDIX C: Banding by Competitive Particle Growth; Practical Software Implementation (Programs PPCPG**).	111
REFERENCES	115
SOFTWARE ORDER FORM	119
INDEX	121

PREFACE

This book represents, not least to the author's own surprise, the third part of his "Crystal Trilogy", the outcome of an enormously enjoyable love affair with crystals and Liesegang Rings. When *Crystal Growth in Gels* (The Pennsylvania State University Press) was published in 1970, it was very much a playful "first book" on the subject, designed to summarize the scattered information, and to entice enterprising researchers into the field. During the following years, the subject flourished, and significant advances were made in terms of applications found and insights gained. This led in 1988 to *Crystals in Gels and Liesegang Rings* (Cambridge University Press), an extensively updated version, the fruit of many additional researches and more mature reflection upon these beautiful phenomena, as observed in nature and in the laboratory. That book also bore witness to a new fact of life: the arrival of the microcomputer on the research scene, though it managed, at the time, to provide only an introductory glimpse of the opportunities ahead. The present book is entirely concerned with microcomputer analysis and software development. It is not an episodic account of laboratory observations, nor is it in any sense a literature survey; it deals, instead, with periodic precipitation phenomena in terms of mathematical models and their logical consequences. In most respects, those consequences are outside the scope of intuitive prediction, just because they are the outcome of complex interactions between the various descriptive parameters of the modelled system. In such circumstances, computer methods and hypothetical experimentation hold the stage.

I should like to thank my good friends, Cornel Popescu and Jean-Claude Manifacier, who first introduced me to computers and alerted me to their enormous potential, as well as Juan-Manuel García-Ruiz, with whose collaboration and encouragement I took the first steps in precipitation modelling. It is also a pleasure to acknowledge Bonny Farmer, who has managed, with great skill and patience over the years, to keep an increasingly absent-minded pilot on course. Lastly, I would like to thank my wife, Bridget, for sharing the adventure, and also for yielding me so generously and so often to the seductive charms of a micro-electronic mistress.

H.K.H.

University Park, Pennsylvania
June 1990

[A form is provided for ordering the associated IBM-compatible software package.].

List of Symbols

The symbols given below are listed in their text-form. When used in BASIC code, they are (as always) printed without subscripts. Thus, for example, the symbol A_{MAX} in the text or in text equations becomes AMAX when used in computer code. The S-number (or equivalent) in curly brackets gives the first Section with which each symbol is explicitly or implicitly associated.

a distance between adjacent jump sites. {S 1.2}

A, $A_0(0)$, $A_1(0)$ boundary concentrations of reagent [A] at $X=0$. {S 1.4}

$A(L)$, $A_0(L)$, $A_1(L)$ boundary concentrations of reagent [A] at $X=L$. {S 1.4}

$A(R)$, $A(X)$ concentration of reagent [A] as a function of the linear distance X or radial distance R. {S 1.2}

$A_0(X)$, $A_1(X)$ concentration of reagent [A] for two iteration stages. {S 1.4}

A_{M0} concentration of [A] with which the medium is pre-charged from $X=0$ to $X=X_{AM}$. {S 5.3}

A_{MAX} maximum concentration of reagent [A]. {S 1.3}

A_{R0} initial reservoir concentration of reagent [A]. {S 4.6}

$B(0)$, $B_0(0)$, $B_1(0)$ boundary concentrations of reagent [B] at $X=0$. {S 2.4}

$B(L)$, $B_0(L)$, $B_1(L)$ boundary concentrations of reagent [B] at $X=L$. {S 2.4}

$B(X)$ concentration of reagent [B] as a function of distance X. {S 2.4}

$B_0(X)$, $B_1(X)$ concentration of reagent [B] for two iteration stages. {S 2.4}

b_A random walk constant, associated with reagent [A]. {S 1.4}

b_B random walk constant associated with reagent [B]. {S.2.4}

B_{R0} initial reservoir concentration of reagent [B]. {4.6}

$C(X)$ concentration of solute [C] as a function of distance X. {S 6.2}

$C_0(X)$, $C_1(X)$ concentration of solute [C] for two iteration stages. {S 6.2}
C_E equilibrium concentration, variable; initial value C_{E0}. {S 6.2}
C_{EM} minimum equilibrium concentration. {S 6.2}
C_{RH} concentration of solute [C] in the "high" C-reservoir. {S 6.2}
C_{RL} concentration of solute [C] in the "low" C-reservoir. {S 6.2}

D (diffusivity) diffusion constant in general. {S 1.2}
D_A (diffusivity) diffusion constant of reagent [A]. {S 1.2}
$D_A(X)$ diffusivity of reagent [A] when non-constant. {S 2.2}
D_{A0} diffusivity of reagent [A] for infinite dilution. {S 1.3}
D_{A1}, D_{A2} particular values of D_A in an non-uniform medium. {S 2.2}
D_B (diffusivity) diffusion constant of reagent [B]. {S 2.4}
D_C (diffusivity) diffusion constant of [C]. {S 6.2}
D_{EC} concentration decrement (general). {S 3.3}
D_{ECG} concentration decrement due to deposit growth. {S 4.1}
D_H (diffusivity) diffusion constant of the secondary reaction product [H]. {S 4.5}
D_{NC} deterministic nucleation coefficient. {S 4.4}

e electric charge of diffusing particle. {S 1.3}
E electric field. {S 1.3}
E_R "equality range", relating to stoichiometry requirements. {S. 3.2}

f jump attempt frequency in basic diffusion theory.. {S 1.2}
F_{AT} (electric) field application time; temporary field. {S 5.8}
F_F index governing whether the electric field should be static, temporary or alternating. {App A}
F_{RT} (electric) field removal time; temporary field. {S 5.8}

G_{RC} growth coefficient. {4.1}

$H(X)$ concentration of the secondary reaction product [H], as a function of X. {S 4.5}
$H_0(X)$, $H_1(X)$ concentration of [H] for two iteration stages. {S 4.5}
H_{IC} index providing a program choice, relating to the interaction with [H]. {S 5.6}
H_{PI} auxiliary variable, relating to the half-period of an alternating square-wave field. {S 5.8}

I_{NCS} concentration increment resulting from re-solution. {S 4.1}

J total current density. {1.3}

J_{DA} diffusion current density associated with the reagent [A]. {1.3}
K rate constant governing competitive particle growth. {S 6.2}
K_S solubility product. {S 3.1}
K_{S0}, K_{SP0} initial values of K_S and K_{SP}, when time-dependent. {S 4.5}
K_{SP} precipitation product. {S 3.1}

L length of the diffusion medium
m_A electrical mobility of [A]-ions. {1.3}

$M_{PG}(X)$ (uniform) mass per grain as a function of X. {S 4.1}
M_{PG0} initial value of the (uniform) mass per grain. {S 3.3}

$N(X)$ number of grains. {S 3.3}
$N_P, N_P(X)$ nucleation probability. {S 3.2}
N_{UC} index providing a program choice relating to the implementation of random nucleation. {S 5.9}

P_{HF} pH-factor which controls the interaction of a secondary product [H] with the primary precipitation product. {S 4.5}
P_{RI} phase reversal index for alternating fields. {App A}

R radial distance. {S 1.3}   particle radius. {S 6.2}
R_0 initial particle radius. {S 6.2}
R_{DC} reservoir depletion coefficient. {S 4.6}
R_{HO} particle density. {S 6.2}
R_R reduced radius; auxiliary variable. {S 3.2}
R_{SC} re-solution coefficient. {S 4.1}
R_T total number of runs. {S 4.4}

S_{AT} saturation concentration. {S 6.1}
S_{EN} sensitivity factor. {S 6.1}
S_{PT} special printout time. {App A}
S_{SA} system status (closed or open) index at the [A]-reservoir boundary. {App A}
S_{SL} supersaturation limit. {S 6.1}
S_{SL0} initial value of S_{SL}. (S 6.1)

T time. {S 1.2}
T_{HP} half-period of the square-wave alternating electric field. {S 5.8}
$T_M(X)$ total deposited mass, as a function of X. {S 4.4}

$T_N(X)$ total number of grains, as a function of X. {S 4.4.}
T_{RD} video updating time (interval). {App A}
T_T total time of a computer run. {S 1.4}

X distance from the boundary of the [A]-reservoir, along the diffusion column. {S 1.2}
X' adjusted (re-normalized) value of X. {S 2.1}
X_{AM} distance to which the diffusion medium is pre-charged with [A], to a concentration A_{M0}. {S 5.3}
X_{PB} X-value at which the particle distribution begins. {S 6.2 and App C}
X_{PE} X-value at which the particle distribution ends. {S 6.2 and App C}

Y auxiliary variable = $X/2(D_A.150)^{1/2}$. {S 2.1}

1. INTRODUCTION

1.1 General Considerations.

This is a dual purpose book. On the one hand, it is concerned with a specific phenomenon, namely periodic precipitation (see below), and with its component mechanisms; on the other, it introduces the reader to some of the procedures of computer analysis, and to the possibilities of research by hypothetical experimentation. In that last sense, the discussions have a mission of a more general character. It is also one that will be of ever-increasing importance in the years to come. In the chapters which follow, the two objectives are necessarily interwoven, but their distinctive characters will always be clear.

Periodic precipitation is a generic term for processes of material deposition which occur intermittently in terms of time or space or (generally) both. Such processes represent a special case of a fashionable topic: oscillatory reactions, with practical implications in geology, crystal growth, and materials preparation, and a theoretical kinship with the complex of problems that come under the heading "order out of chaos" (e.g. see Prigogine, 1984). The term "periodic" is actually a piece of literary license, and should be interpreted with caution and tolerance. Strictly speaking, it demands a period, i.e. a constant time interval, something that the phenomena here under discussion fail to show. These phenomena are nevertheless "periodic", in the the same sense in which (say) a frequency-modulated wave can be described as periodic. The periods of such a wave are not constant, but they are not random either, and vary in response to a control process, singular and simple in the case of FM, multi-component and complex here.

Periodic precipitations are often referred to as "Liesegang Rings", because, in ring form (which is only one of their many manifestations), they were first extensively described by Liesegang (1896, 1897, 1898). Fig. 1.1.1 gives examples drawn from the work of Hatschek (1914), and shows lead iodide rings. Since the days of Liesegang, such systems have been the subject of many researches, of which a comprehensive (and, one is tempted to say, masterly) overview has already been presented in previous books; see Henisch 1970 and 1988, and the bibliographies given there. The 1988 book also contains a brief biography of R.E.Liesegang, an interesting man and formidable scientist by any reckoning.

Liesegang Rings can manifest themselves in a great variety of media and ways, but the essentials are simple enough. Reagents, whether in nature or in the laboratory, diffuse in some (preferably inert) environment, and in due course react with one another to produce spaced deposits, each consisting of insoluble (or only sparingly soluble) clusters or crystals. Fig. 1.1.2a gives an example of banded silver chromate deposits, resulting from the diffusion of silver nitrate into a silica gel charged with chromic acid. In this case, the crystallites are relatively large, and their number and size can be seen to vary from layer to layer. In other cases, the deposits may be microcrystalline or amorphous, and the layer systems themselves much more intricate than those shown. Fig. 1.1.2b represents a particularly well-developed system of microcrystalline calcium phosphate in a tube, and incidentally illustrates some radial structure effects which have so far defied analysis. Yet another manifestation is shown in Fig. 1.1.3, drawn from the realm of biochemistry. The striations are due to metallic silver, deposited in the leaf epidermis of the halophytic plant, *Ceratostigma plumbaginoides* (Borchert, 1989).

For laboratory research, silica gel turns out to be a particularly suitable diffusion medium, because of its mechanical flexibility and chemical inertness, but in practice periodic deposits have been observed in many other media, from sand (in nature) to agar, alumina gels and solid copper (Gerrard et al., 1962). Three kinds will be described: (1) deposits arising from chemical reactions between component species [A] and [B], forming an insoluble (or only sparingly soluble) compound [AB] of concentration AB, (2) deposits arising from a single component (monomer), as a result of solubility modulation, and (3) monomer deposits resulting from competitive particle growth.

The phenomenology of Liesegang Ring formation, in all its varied contexts and manifestations, has already been widely discussed, and is therefore outside the present scope. The present task is generic analysis. Accordingly, we will not be concerned with specific diffusion media, specific diffusants or specific precipitates. Instead, we will focus on "systems", characterized by descriptive "parameters", with the way these parameters interact to produce different kinds of layered systems, and with the possibilities of controlling the structure of such systems, e.g. for the purpose of making anisotropic ceramics. (Adair et al., 1987). In the computer (as, indeed, in real life) the chosen parameters determine the

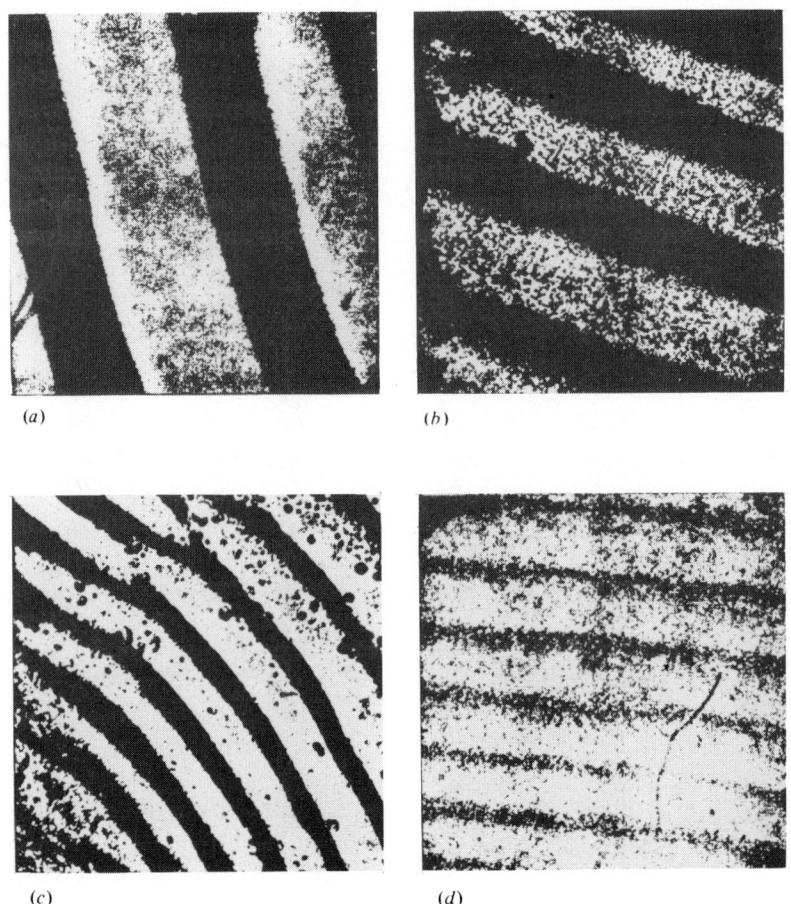

Fig. 1.1.1. Concentric ring systems in agar, (a) silver chromate (wet) x100, (b) lead iodide (wet) x100, (c) lead iodide after drying, x50, and (d) lead chromate (wet) x50. After Hatschek (1914).

outcome. Because there are many such relevant parameters, and because they interact in complicated ways, the logical consequences of any particular combination of chosen descriptors cannot, in general, be intuitively predicted, and the possibilities of doing genuine research by hypothetical computer experimentation derive from that central fact. The arrival of the microcomputer brings these possibilities to virtually any scientist who is tempted to take advantage of them.

4   Periodic Precipitation

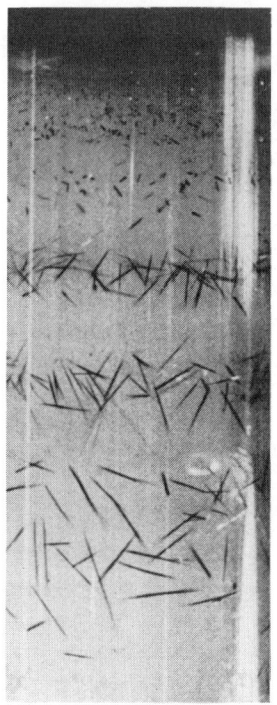 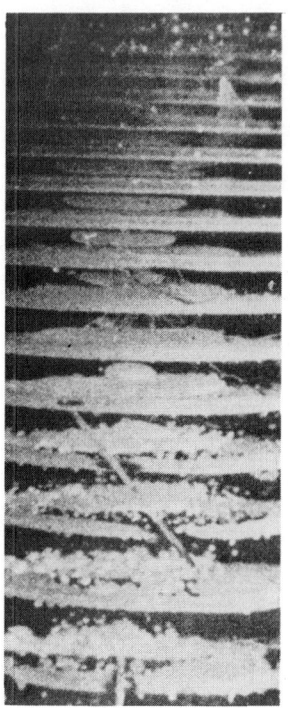

Fig. 1.1.2. Liesegang ring systems of (a) silver chromate, and (b) calcium phosphate. After Henisch (1988).

Of course, the results of analysis must in due course be compared with actual observations made under a variety of conditions. As a rule, initial agreements are not hard to obtain (the starting model is always formulated to guarantee that much), but a time is bound to come when the two sets of data begin to diverge. In such circumstances, a traditional theorist turns to a new *theory*, or else improves his old one. In a similar way, a computer-theorist turns to a new *algorithm* or makes his old one more sophisticated. Both do it in the hope that the activity will yield deeper insights and, indeed, science owes its success to the fact that it often does. One could say that when observational data are available for comparison purposes, then we do research *on* the algorithm; when they are not available, we do research *with* it, and formulate predictions for future experimental tests. There are many fields in which traditional

Introduction   5

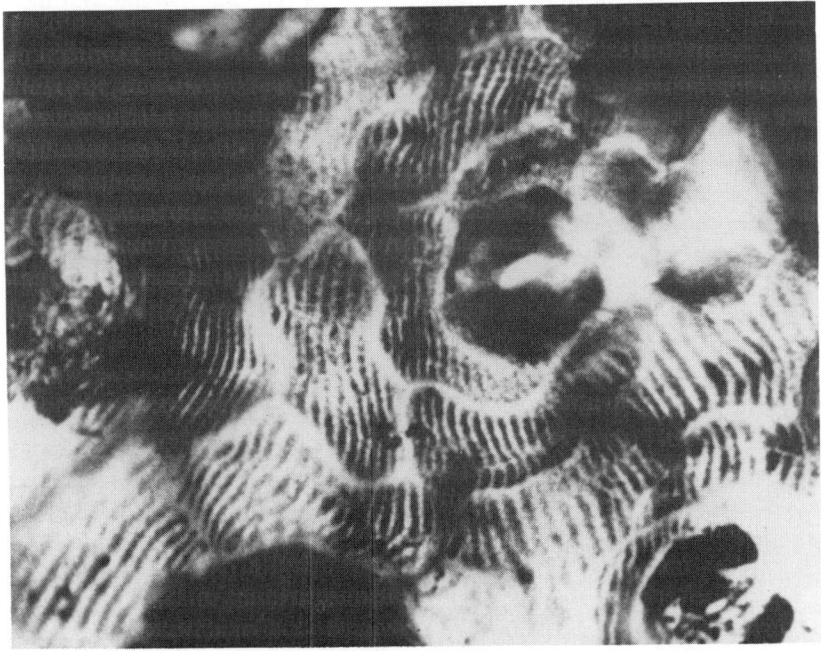

Fig. 1.1.3. Periodic precipitation of metallic silver in the leaf epidermis of a halophytic plant. The large epidermis cells contain a viscous muticilage, consisting of acidic polysaccharides and high salt concentrations. The deposits form after immersion of the leaf in a 5% solution of silver nitrate. Contributed by R. Borchert. See also Borchert (1989).

(analytic) and computational methods compete with one another, but in the specific context of periodic precipitation, they actually don't; the situations described are mostly too complicated for analytic procedures to have any chances of success. Here, in a way that would have been inconceivable to early precipitation pioneers, computer methods reign supreme. Not only that, there is an important bonus: the ready linkage between computation and visual display. That permits us to watch "phenomena" on a monitor, as they "happen", and to gain insights from the relationships between them.

For an interesting essay on computer-modeling in general, see Blackman (1987). A solid overview of simulation methods in great variety has been given by Gould and Tobochnik (1988) and, at this point, a glowing tribute should also be paid to a classic, Bennet's (1976) pioneering book on microcomputer problem solving. In the last analysis, computer modelling, albeit a science, retains the hallmarks of an art. Some of its steps follow compelling logic, others depend on facts firmly believed in, others still follow hunches and improvisations, designed to bridge the gap between what is and what might be. Therein lies much of its charm.

The chapters which follow will deal first with diffusion patterns as such, then with chemical reactions between diffusants, precipitation, precipitate growth and re-solution, examples of hypothetical experimentation and, ultimately, with new findings based on the developed algorithms. As everyone knows, computerists use many different languages in the course of their work, which creates the problem of finding a common tongue. In the face of this problem, it is normal practice to describe algorithms in terms of a "pseudo" language, one that does not actually *run* on *any* computer, but is supposed to be so transparently simple and logical as to be instantly understood by all practitioners. Alas, this sensible hope is not always fulfilled. The language used in the following chapters is True BASIC, invented (as, indeed, was the original BASIC) by John G. Kemeny and Thomas E. Kurtz. True BASIC is indeed a very real language (fast, because run in compiled form), and the fact that it is highly structured, elegant, and totally free of "gimmicks" that have to be laboriously learned also make it very suitable for use as a pseudo-language. Another of its cardinal virtues is that its code (albeit not on the same disks!) will run on Macintosh as well as IBM-compatible machines. The language will here serve to formulate the key elements of the algorithm, and show how the various adjustable parameters enter into the scheme of things, but the development of complete, runable programs is not one of the present tasks, nor indeed is it to teach readers the language as such. Nevertheless, Appendices A, B. and C describe in some detail how practical implementations of the programs might be put together.

A computer is not actually needed by readers of this text, but the availability of such a machine (preferably with an EGA or VGA color display) will add to the enjoyment. Of course, a computer *is* needed if the algorithms and procedures discussed are to be used for research. For this purpose, a program disk* (with extensive graphics implementations)

* **Periodic Precipitation Software**, obtainable from The Carnation Press, P.O.Box 101, State College, Pa. 16804. See Appendices A-C and inserted order form for details.

is available, containing the necessary software in compiled, self-supporting form. There is no need to learn any language, since the programs are self-prompting and largely self-explanatory.

The reader can simply regard this text as an account of a successful programing episode, an exercise in microcomputer use for problem solving. Alternatively, the reader could make use of the associated software for hypothetical experiments that extend and advance the examples presented in Chapter 5. Yet another possibility is to design a new (and improved!) program, based on the present experience, and on the reader's special needs and interests. In any event, the author's best wishes for success and enjoyment accompany this book.

1.2 Fick's Laws of Diffusion.

Diffusion Theory a specialized subject, on which many treatises have been written, and its ideas are intimately linked to the concept of the "random walk". One considers particles at certain sites in a medium, particles that can jump to adjacent sites in any direction, and do so as a result of thermal agitation. There is a characteristic "attempt frequency", but whether jumps are successful depends on whether adjacent sites are empty or already occupied. Exactly what constitutes a "site" is a separate question, one that is governed by the nature of the diffusion medium. We will here consider that the medium is isotropic, which means that the probability of a completed jump will not be systematically influenced by its direction. However, it will also be shown that the superposition of an electric field can make the situation (if not the medium itself) anisotropic.

In a homogeneously populated medium, the chances of a jump to the left and the right are equal, which means that there is no net mass transport (flux). Conversely, when concentration gradients are present, then there is such a flux in the direction of the gradient. Indeed, in the absence of electric fields, concentration gradients are the sole driving "forces" behind all the phenomena. A fine exposition of the subject has been provided by Ghez (1988), who starts from first principles, and derives the familiar diffusion equation (also known as Fick's Second Law):

$$\delta A/\delta T = D_A(\delta^2 A/\delta X^2) \qquad (1.2.1)$$

8   Periodic Precipitation

for the one-dimensional case. Here, A = concentration of the diffusant [A], T = time , X = distance. D_A is a constant which is characteristic of the medium and also of [A], but will be regarded, in the first instance, to be independent of the concentration A itself; it is called the *diffusion constant*, or *diffusivity*. In the course of this derivation, Ghez gives clear insight into the assumptions and approximations involved. This is highly desirable, because the basic diffusion equation is often accepted as gospel, whereas it is in fact very much dependent on the simplifications introduced. Indeed, those are responsible for making D_A *constant*, a matter which is often taken for granted. It can be shown that D_A (or, in general, D) is given by:

$$D = f.a^2/2 \qquad (1.2.2)$$

where f is the attempt frequency, and a the distance between adjacent jump sites. The frequency is governed by the mass and size of the jumping particle, and by the temperature, while the jump distance relates D to the structure of the diffusion medium. Different diffusants, say [A] and [B], are in general associated with different diffusion constants, D_A and D_B.

Equation (1.2.1) actually follows from a simpler one, namely Fick's First Law, which relates the rate of mass transport (flux) to the concentration gradient under the (obviously necessary) condition of flux continuity:

$$\text{Flux} = -D_A(dA/dX) \qquad (1.2.3)$$

Because the diffusion equations (1.2.1 and 1.2.3) are of great importance in a variety of contexts (including, for instance, charge transport in semiconductors), their properties are by now very well established, in so far as they can be obtained by analytic (algebraic) methods, but it is also true that such methods work only in rather simple circumstances, e.g. diffusion across homogeneous membranes, with fixed boundary conditions; see Crank (1956), Shewmon (1963), Rainville (1963), and Ghez (1988). In contrast, the situations with which we are here involved call for numerical (computer) solutions.

1.3 Modified Diffusion Equations.

It is convenient at this stage to consider how the above diffusion equations would have to be modified if the system were just a little less simple than so far envisaged. One might, for instance, encounter a diffusion medium that is structurally non-uniform from the beginning, or else one that *becomes* non-uniform in the course of mass precipitation within it. As a result, the diffusion constant would become a function of the space coordinate, leading to:

$$\frac{\delta A}{\delta X} = \frac{\delta}{\delta X}\left[D_A \frac{\delta A}{\delta X}\right] = D_A \frac{\delta^2 A}{\delta X^2} + \frac{\delta D_A}{\delta X} \frac{\delta A}{\delta X} \quad (1.3.1)$$

Then there is the (independent) question of whether D_A depends on A. Equation (1.2.1) assumes implicitly that there is never an occupancy problem; if a particle can jump to a neighboring site, it will find that site unoccupied or, at any rate, able to accept another occupant. The situation would be very different if the occupancy were limited, in the extreme case, to a single particle. The first consequence would be to establish a maximum value for A, say A_{max}, corresponding to total occupancy of all sites. The second consequence would be to limit the probability of making a successful jump, and this matter has also been investigated (see Ghez and Langlois, 1986). For present purposes, a crude, intuitive approximation will do, because we will be concerned only with qualitative onsequences of concentration effects. We shall therefore write, for the case of single particle occupancy:

$$D_A(A) \approx D_{A0} \cdot (1 - A/A_{max}) \quad (1.3.2)$$

with $0 < A < A_{max}$, where D_{A0} is evidently the diffusivity for infinite dilution. If the diffusion medium were (say) a silica gel, then concentration effects would be very unlikely, because the pores tend to be orders of magnitude larger than the sizes of diffusing species. On the other hand, if we were dealing with diffusion in a lattice (with defect vacancies), then the situation would be very different, and concentration effects would have to be very carefully explored. In most contexts, it will here be assumed that they can be neglected, but a brief exploration of their consequences will be found in Section 2.2.

Special considerations apply when the particles which constitute the diffusant [A] are charged. If each particle were to carry a charge $-e$, then equ.(1.2.3) would imply a diffusion current density J_{DA}, given by:

$$J_{DA} = eD_A \cdot (\delta A/\delta X) \quad (1.3.3)$$

even in the absence of an electric field. [Quick sign check: With A increasing towards the right, the particles would diffuse towards the left. Since they (here) carry a negative charge, that movement would amount to a positive current.] If an electric field E were superimposed on the concentration gradient then the total current density J would be the sum of the diffusion and field components:

$$J = eD_A.(\delta A/dX) + em_A A.E \qquad (1.3.4)$$

where m_A is the electrical *mobility* of the [A] particles. Since we are here concerned not with currents but simply with particle fluxes, the expression which corresponds to equ.(1.2.1) becomes:

$$\frac{\delta A}{\delta T} = D_A \frac{\delta^2 A}{\delta X^2} + m_A E \frac{\delta A}{\delta X} + m_A A \frac{\delta E}{\delta X} \qquad (1.3.5)$$

To solve the problem with E as a second variable, one would have to make use of a second relationship, namely Poisson's equation, which relates the local charge density to the local field gradient. One component of the field E would be produced by the varying concentration A itself, since each local concentration would correspond to a local charge density. A second component could be imposed by external means, that is to say by the application of some voltage between parallel plates. For the sake of simplicity, we will here assume that this second component is far larger than the first, which means that the field can be regarded as uniform. The field-modified diffusion equation for negative charges then becomes:

$$\frac{\delta A}{\delta T} = D_A \frac{\delta^2 A}{\delta X^2} + m_A E \frac{\delta A}{\delta X} \qquad (1.3.6)$$

and it is in this form that it will here be used (see Sections 2.3, 2.4 and 5.8). The necessity of using Poisson's equation is thereby avoided, but the limitations involved must not be forgotten.

Lastly, there is the question of the most appropriate coordinate system. Since precipitates are nucleated in space, a three-dimensional coordinate system would certainly be the most realistic, but it would also be prohibitive in terms of (micro)-computing power. The computations will therefore be performed in one dimension, with linear coordinates. For cylindrical coordinates, equation (1.2.1) would have to be written as:

$$\frac{\delta A}{\delta T} = \frac{DA}{R} \left[R \frac{\delta^2 A}{\delta R^2} + \frac{\delta A}{\delta R} \right] \qquad (1.3.7)$$

where R is the radial distance. This makes analytical solutions more difficult, but the computer would take it in its stride.

Of course, there is a price to be paid for these simplifications. For one thing, we will not be able to describe the growth of *individual* crystallites or clusters within a population. The average size of these clusters in a precipitate may (and generally does) vary with X, but at any give value of X the particles will be assumed to be spherical (a common physicist's dream) and of uniform size. A glance at Fig. 1.1.2 will show that this is not wrong as far as size is concerned, but the sphericity of those silver chromate crystal leaves something to be desired.

1.4 Basic Diffusion Algorithms

Section 1.3 deals with continuum relationships. For computational purposes, these have to be converted to finite difference formats. The procedures for doing so are standard and well-known. See, for instance, Bajpai *et al.* (1977). Accordingly, we consider in the simplest case:

$$\frac{\delta A}{\delta T} = \frac{A(X+\Delta X, T) - A(X, T)}{\Delta T} \tag{1.4.1}$$

but it is sometimes more appropriate to write:

$$\frac{\delta A}{\delta T} = \frac{A(X + \Delta X, T) - A(X - \Delta X, T)}{2\Delta T} \tag{1.4.2}$$

for the sake of symmetry. In the same mode, we also have:

$$\frac{\delta^2 A}{\delta T^2} = \frac{A(X + \Delta X, T) - 2A(X,T) + A(X - \Delta X, T)}{\Delta X^2} \tag{1.4.3}$$

In such terms, equation 1.2.1 becomes, after substitution and re-arrangement:

$$A(X, T + \Delta T) = \frac{\Delta T D_A}{\Delta X^2} A(X - \Delta X, T) + \left[1 - \frac{2\Delta T D_A}{\Delta X^2}\right] A(X,T)$$

$$+ \frac{\Delta T D_A}{\Delta X^2} A(X + \Delta X, T) \tag{1.4.4}$$

or

$$A(X, T+\Delta T) = b_A \cdot A(X-\Delta X, T) + (1-2b_A) \cdot A(X,T) + b_A \cdot A(X+\Delta X, T) \tag{1.4.5a}$$

where $$b_A = \Delta T . D_A / \Delta X^2 \tag{1.4.5b}$$

The "random walk" character of b_A will be evident. It can be shown that random walk processes are characterized by a *diffusion length* given by $(D_A T)^{1/2}$, and the numerator of equation 1.4.5b thus has the dimension of (length)^2, making b_A dimensionless. Of course, the remaining variables have their ordinary dimensions, but the fact is that computers do not understand those; they understand only numbers. One way out would be to normalize all the variables explicitly, in order to arrive at a dimensionless form of the equation. This could certainly be done, but it is even simpler (if not mathematically more virtuous) to *assume* that it has already be done. Thus, every variable will be deemed, from now on, and unless otherwise stated, to have been normalized to some (unspecified) reference value, and thus dimensionless. This is in harmony with the central objective, which is not to compute highly specific real-world cases, but to elucidate relationships and general modes of behavior.

Having done that, we can actually make a further simplification with impunity, by making $X = 1$ and $T = 1$. This makes the b_A parameter numerically equal to D_A, the diffusion constant of [A]. In the circumstances, there is no longer any need to distinguish between these variables in the computational equations. We therefore write:

$$A(X,T+1) = D_A A(X-1,T) + (1-2D_A).A(X,T) + D_A A(X+1,T) \qquad 1.4.6$$

This means that if we know the value of A at three points, namely at $X - 1$, X, and $X + 1$, at a time T, we can calculate what the next value $A(X)$ will be, *next* meaning at a time $T + \Delta T = T + 1$.

It can be shown that b_A should be less than 1/2 for stability (see Milne, 1953), and also that the computational accuracy is greatest when $b_A = D_A = 1/6$. Of course, there is always a question of how much accuracy is enough, but we shall accept the value of 1/6 for the moment, and will later examine the importance of this constraint. Equation 1.4.6 then becomes

$$A(X, T+1) = [A(X-1,T) + 4A(X,T) + A(X+1,T)]/6 \tag{1.4.7}$$

which has a straightforward interpretation: the new A value at X is simply the weighted average of the three adjacent values at time T.

We can now perform the same operation on equations 1.3.1, 1.3.6, and 1.3.7. Thus, beginning with equation 1.3.6, we can write, this time for *positively* charged particles:

$$A(X, T+\Delta T) = (1/6).[A(X-\Delta X,T) + 4A(X,T) + A(X-\Delta X,T)]$$

$$+ m_A Ee.[\Delta T/2\Delta X].[A(X-\Delta X,T) - A(X+\Delta X,T)] \quad (1.4.8)$$

the last term being once again formulated for a double interval in order to achieve symmetry. We see that the field contribution depends not only on the sign of E, but also on the sign of the concentration gradient. With $\Delta X=1$, $\Delta T=1$, $D_A = 1/6$, as before, this gives

$$A(X,T+1) = \left[\frac{1}{6} + \frac{m_A eE}{2}\right].A(X-1,T) + \frac{4}{6}A(X,T)$$

$$+ \left[\frac{1}{6} - \frac{m_A eE}{2}\right].A(X+1,T) \quad (1.4.9)$$

The mobility m_A is actually proportional to the diffusion constant D_A, as long as concentration effects are absent, and this proportionality is well-known as *Einstein's Relationship*. Thus, D_A is not an independent variable. Moreover, we are at this stage committed to regarding E as normalized, but have not yet said *to what*. These facts make it possible for us to "re-define" E by absorbing all constants of proportionality and writing (with carefree abandon):

$$A(X,T+1) = D_A(1+D_A.E).A(X-1,T) + (1-2D_A)A(X,T)$$

$$+ D_A(1-D_A.E).A(X+1,T) \quad (1.4.10)$$

whereby the charge of the diffusing particles has again been taken as positive; compare equation 1.4.7. At the same time, allowance has been made for varying the diffusion constant. The electric field thus affects the nature of the averaging process over the three adjacent A-values, and does so in a polarity dependent way, as one would expect.

The transformation of equation 1.3.1 (for a non-constant D_A) can proceed in a similar way, when D_A is given as a function of X. Maintaining symmetry, equation 1.3.1 then yields equation (1.4.11) in a form in which it will be used in Section 2.2. It clearly reduces to the root form when D_A is constant.

14 Periodic Precipitation

$$A1(0) = A(X-1).[D_A(X)-D_A(X+1)/4+D_A(X-1)/4]$$
$$+ A(X).[1-2D_A(X)]$$
$$+A(X+1).[D_A(X)+D_A(X+1)/4-D_A(X-1)/4 \quad (1.4.11)$$

and in this form it will be used in Section 2.2. It will be seen that this reduces to the root form when D_A is constant.

At other times, D_A might be given as a function of concentration, as in equation 1.3.2, in which case the second term of equation 1.3.1 becomes:

$$\frac{\delta D_A}{\delta X}\frac{\delta A}{\delta X} = \frac{\delta D_A}{\delta A}\left(\frac{\delta A}{\delta X}\right)^2 = -\frac{D_{A0}}{A_{max}}\left(\frac{\delta A}{\delta X}\right)^2 \quad (1.4.12)$$

which includes the schematic approximation for a concentration-dependent diffusion constant represented by equation 1.3.2. Thus, for $D_{A0} = 1/6$,

$$A(X,T+1) = [A(X-1,T) + 4A(X,T) + A(X+1,T)]/6$$
$$- [A(X-1,T)^2-2A(X-1,T).A(X+1,T)+ A(X+1,T)^2]/6A_{max} \quad (1.4.13)$$

A similar treatment of equation 1.3.7, for radial geometry, leads to:

$$A(R,T+1) = \frac{1}{6}\left[1 - \frac{1}{2R}\right]A(R - \Delta R,T)$$
$$+ \frac{2}{3}A(R,T)$$
$$+ \frac{1}{6}\left[1 + \frac{1}{2R}\right]A(R+\Delta R,T) \quad (1.4.14)$$

the radial distance R having been normalized. No surprises are in store for us on this account. When R is large, the $1/2R$ term tends to be unimportant, and the whole expression approaches its linear form. When R is small, concentration gradients are higher than those in the linear case. Radial systems are rarely used for quantitative research, and are included here, in passing, only for the sake of completeness.

In order to illustrate how these algorithms are actually implemented, we return to the simplest form, namely equation 1.4.7. Let one-dimensional diffusion medium consist of L-1 cells, each of length X (which we have already set to unity). L is then also the normalized length of the medium. The locations $X=0$ and $X=L$ correspond to the reagent reservoirs on each side of the medium, making $A(0)$ and $A(L)$ the boundary concentrations at time $T = 0$. See Fig. 4.6.1. The two values would be determined by user input. We shall (within the framework of this example) assume that the reservoir of diffusant [A], the only diffusant in this case, is inexhaustible, i.e independent of time, at the level $A(0) = 100$. For the moment it will also be assumed that the diffusion medium is initially empty: $A(X) = 0$ for $X>0$ at $T = 0$.

When the updating procedure is actually written in code, it is convenient to envisage $A(X)$ simply as a one-dimensional array, rather than a two-dimensional one, as suggested by the above equations. In any event, we neither need nor want to keep every concentration value in memory for every value of T. It is therefore more appropriate to denote the existing array values at time T as $A_0(X)$, and the new, updated values (at time $T+1$) as $A_1(X)$. Of course, those just recalculated values become the "existing values" in the next step, and themselves candidates for updating during the *next* pass. That is why the arrays have to be re-set, as shown below.

[N.B. When equations are intended to be parsed and understood in accordance with the rules of algebra, they are written in terms of normal, italicized variables, with conventional subscripts, where appropriate. When they are intended to be interpreted as *computer code*, they will be printed in a distinctive typeface, without subscripts.]

```
for T = 1 to TT
  for X = 1 to L-1
  let A1(X) = [A0(X-1)+4*A0(X)+A0(X+1)]/6
  next X

  for X = 1 to L-1
  let A0(X) = A1(X)
  next X
next T
```

Here, TT is the total time envisaged for a run. The boundary concentrations at $X=0$ and $X=L$ remain unchanged during these passes. After each pass, the computer program can ask questions about the outcome (e.g. are the conditions for nucleation satisfied?), and take measures to interact with the proceedings.

[N.B. The term "let", used above, is, of course, the traditional assignment command in Basic, though many dialects of the language have since done away with it. True BASIC can also be run in a "nolet" mode, but is more often run, as here, with "let" retained. The term is harmless enough, and serves the purpose of drawing attention to the nature of the code line.]

For most purposes one might envisage a minimum value of $L=50$ for the number of cells. Certainly, accuracy tends to suffer when lower values are chosen, but on occasions when the concentration profile is known to be very simple (e.g. see below), even $L=20$ can be satisfactory. When it is highly convoluted, a much larger number of cells would be required. There is no practical maximum value, save that determined by our patience; the computing time increases rapidly with L.

The question of *when* values should be printed out and in what increments of X is, of course, optional and depends in practice also on the degree of convolution. After the final pass, one might, for instance, want to print every ith value, by simply adding to the above:

```
for X = 0 to L
   if X/i = INT(X/i) then print X, A(X)
next X.
```

2. CONCENTRATION PROFILES

The relationships discussed in this Chapter illustrate the behavior of the diffusion equation, and are designed to build confidence in the computational algorithms employed. For the sake of simplicity, the reservoir concentrations $A(0)$, $A(L)$, $B(0)$ and $B(L)$ will here (but not necessarily elsewhere) be taken as constant, i.e. not subject to depletion or contamination.

2.1 Tests of the Simplest Diffusion Algorithm.

We are now in a position to test the diffusion algorithms so far developed, in the first instance equation 1.4.7, and then equation 1.4.6, with different values of D_A. To do this, we shall assume that species [A] is diffusing into a slab of length $L=20$ (no units, everything is normalized) from the left, from a reservoir of concentration $A0(0)=100$. $A0(L)$ will be taken as zero.

Profiles in the Transient State.

As long as T is small, the medium "looks" to the diffusant as if it were semi-infinite. In such circumstances, we have an analytic solution in terms of the error function, namely:

$$A(X,150) = A(0,150).[1 - \mathrm{erf}(Y)]$$

$$\text{where } Y = X/2(D_A.150)^{1/2}$$

(2.1.1)

and with that solution the computed results can be readily compared; see Fig. 2.1.1a. The agreement for $T=150$ is good to one part in several thousand, more than we ordinarily need. Also shown in the figure are computed results for smaller values of T. Longer times cannot be safely compared in this way, because the finite thickness of the diffusion medium would make itself felt. However, see below.

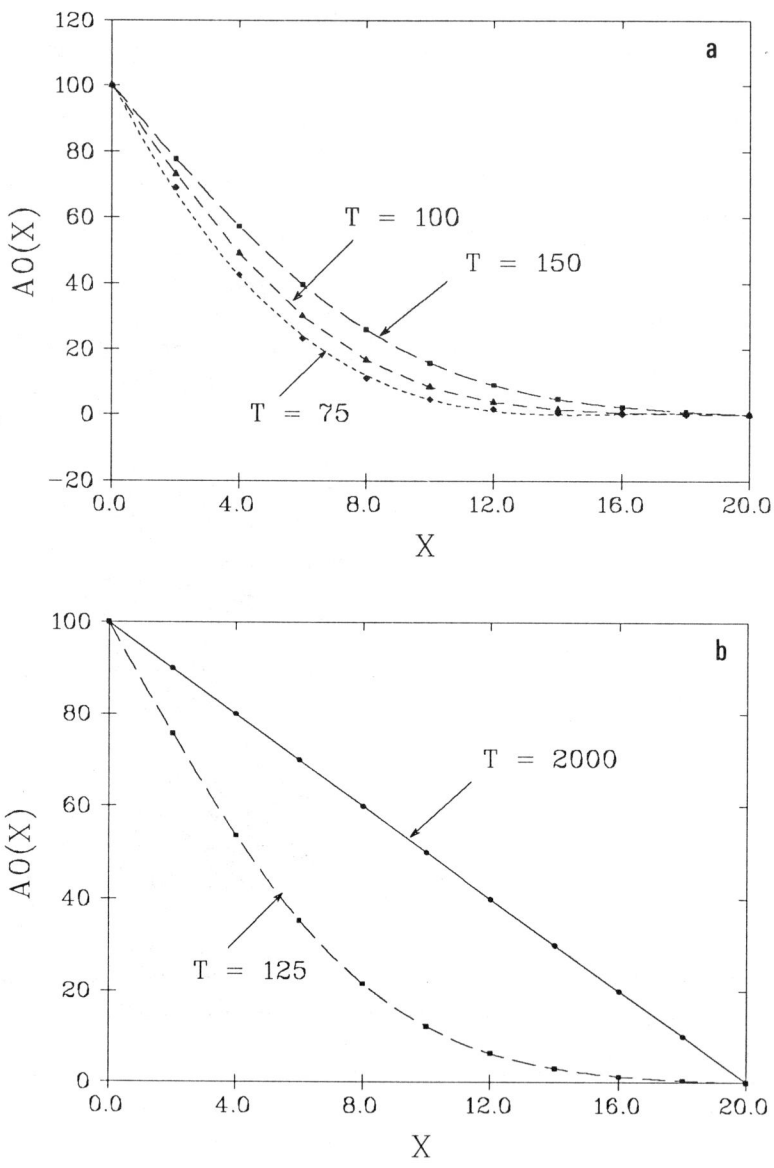

Fig. 2.1.1. Tests of the simplest diffusion algorithm. $L=20$.
(a) Computed results for $T=75$, 100, 150 (small dots). Analytically calculated results for $T=150$: upper curve. Note excellent agreement.
(b) Computed results for $T=125$ and $T=2000$. Note satisfactory approach to linearity.

Profiles in the final state.

When the diffusion process (with time-independent boundary conditions, as here) has been going on for a long time, the concentration profile must reach stability. Equation 1.2.1 (with the left-hand side zero) then implies a linear profile, and this gives us an opportunity for a further test. Do the computations appreach this linearity with increasing T? They do, indeed. Fig. 2.1.1b gives computed results for $T=2000$, which reflect the linearity with very satisfactory accuracy (one part in 10^4).

One concludes that, at any rate for such simple profiles, the algorithm works very well, even when L is as low as 20. For more convoluted profiles, such a small number of cells would be very unsafe.

Range of permissible diffusion constants.

The next test concerns the significance of the $D_A=1/6$ choice, referred to in Section 1.4.3. For $L=20$ and $T=150$, we shall compute concentration profiles for $D_A=0.2$, 0.16667 ($=1/6$), 0.1, and 0.001.

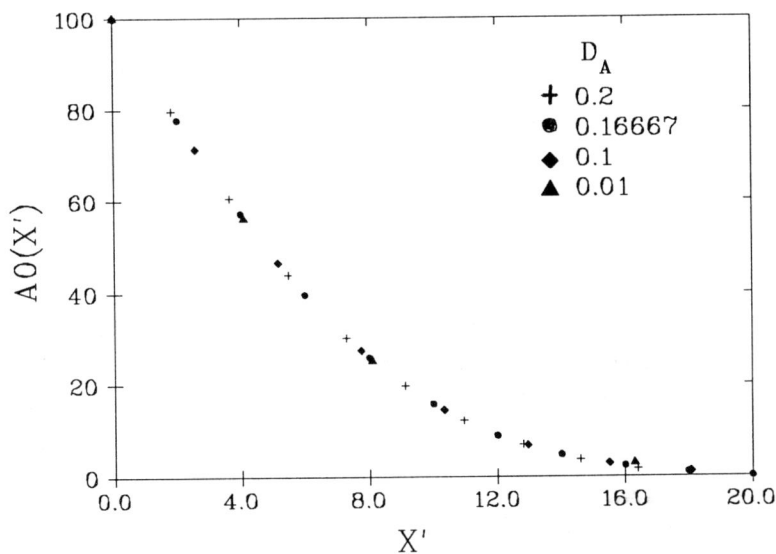

Fig. 2.1.2. Tests for a usable range of diffusion constant values. $L=20$. X' is here the adjusted value of X.

20   Periodic Precipitation

and compare them with results calculated by means of equation 2.1.1. To do this, it is, of course, necessary to correct the abscissa for the different values of D_A implied. Fig. 2.1.2 shows the results. The corrected distances are called X', to distinguish them from the X in previous figures, which applied specifically to $D_A = 1/6$. We see that the agreement is highly satisfactory between $D_A = 0.2$ and $D_A = 0.01$. This gives us a 20:1 ratio of diffusion constants, which should be ample for the present "experimental" purposes. However, these tests are not mathematically rigorous, and we do not actually know when to expect a breakdown of the agreement. On these grounds, caution will prevail, and D_A-variations will be kept within tighter limits.

2.2 Diffusion in Non-homogeneous Media.

Let the length L of the diffusion system be divided into three parts, which are for some structural reason associated with high, low, and high values of $D_A(X)$ respectively, each value being constant within its region. Equation 1.4.11 is appropriately structured for dealing with this case. All that needs to be done is to populate the $D_A(X)$ matrix with suitable values. Thus, for the example shown in Fig. 2.2.1, for which $L=24$, one would program the following lines as part of the set-up procedure. In Basic:

```
for X = 0 to 10
let DA(X) = 0.2
next X

for X = 11 to 16
let DA(X) = 0.05
next X

for X = 17 to 24
let DA(X) = 0.2
next X
```

One can see the effect most distinctly at high values of T, for which the concentration contours in the uniform regions are linear; compare Fig. 2.1.1b.

The results are qualitatively in accordance with expectations. Since the flux must be continuous, the concentration gradient of [A] must be greater in the regions in which $D_A(X)$ is smaller. Indeed, it is, but in the

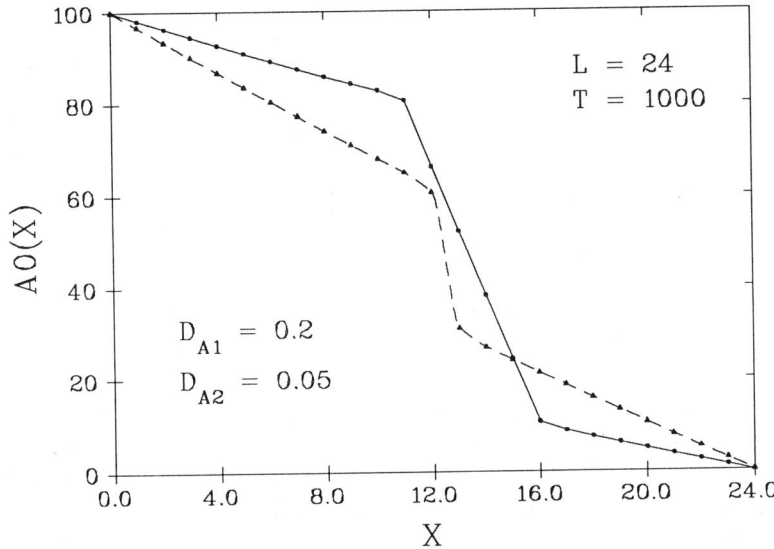

Fig. 2.2.1. Effect of regionally variant diffusion constants on the concentration profile. Low mobility region (narrow: broken line, wide, solid line) between two high mobility regions. $L=24$, $T=1000$.

quantitative realm, Fig. 2.2.1 is actually deceptive. In this example, the diffusion constants vary by a factor of 4, and so should the concentration gradients in the respective regions, but a glance shows that the ratio of slopes is much too high. Great caution is therefore in order. Two disturbing facts are here at work. In the first instance, $L=24$ is not a sufficiently large number to yield the reliable gradients which are needed for this calculation. In the second, the high D_{A1}/D_{A2} ratio means that D_{A2} is very small which, in turn, means that the system takes a long time to equilibriate, indeed, longer than the $T=1000$ here envisaged.

What actually happens can been seen much better in Fig. 2.2.2a, which applies to $L=100$, and a two-layer system divided in the middle into a layer of $D_{A1} = 0.2$ and one of $D_{A2} = 0.05$. One could think of the latter as a dense medium, and of the former as less dense. The curve for $T = 10.000$ is particularly instructive. In the less dense region, there is actually a slight pile-up of [A] near the internal boundary, and in the dense region the profile is (even after that long time) still far from

22 Periodic Precipitation

linear. A later (and, indeed, final) profile is shown by the full line; by that time ($T=20.000$) the slopes are correct.

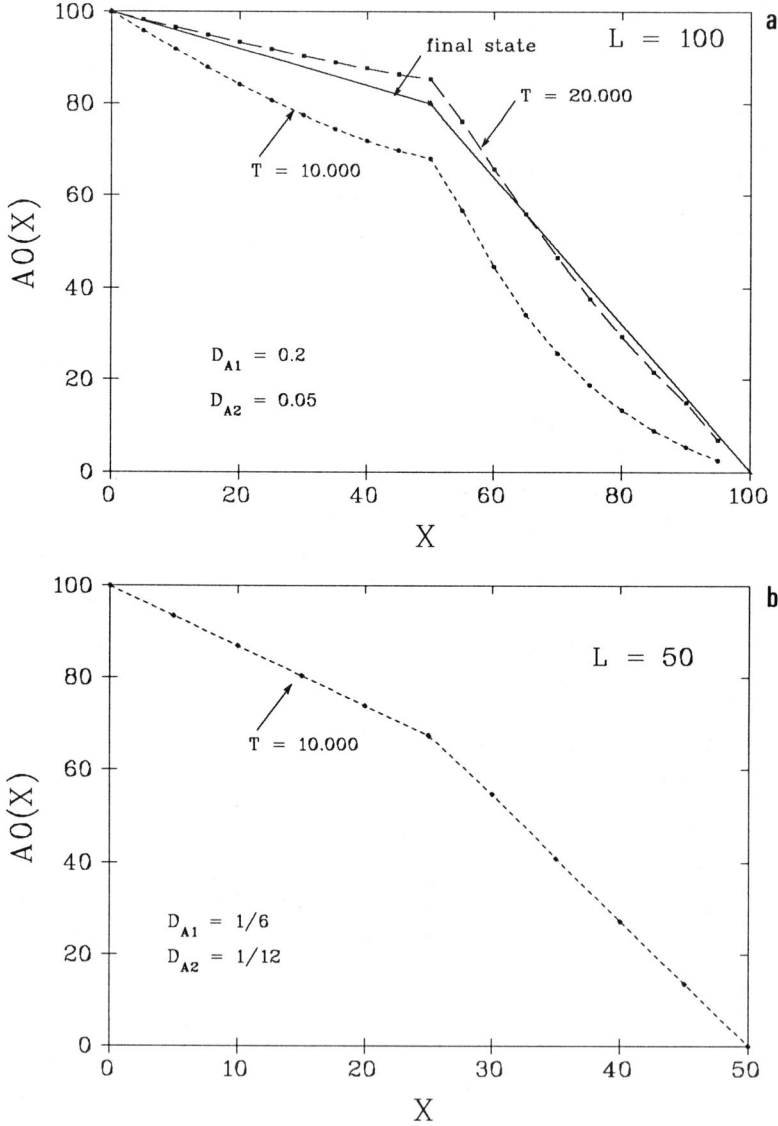

Fig. 2.2.2. Diffusion in non-homogeneous structures; two layer system. (a) high (4:1) ratio of diffusion constants, $L=100$, (b) low (2:1) ratio, $L=50$.

The fact that the algorithm is working properly is also shown by the results in Fig. 2.2.2b, for L is smaller, but the ratio of diffusion constants is smaller also: $D_{A1} = 1/6$ and $D_{A2} = 1/12$. Under these conditions the agreement is already near-perfect at $T=10.000$. Both regions are linear, and the slopes are in the expected ratio of 1:2.

In such non-homogeneous structures, the calculated change-over from one regime to another is re-assuringly abrupt, but this is true only when each region is relatively wide, since each computation of *A1* involves three adjacent cells (values of *X*). Thus, if we had a "low diffusion slab" that is only one cell thick, it would it would leave (with the above D_A-values) practically no mark on the resulting contours.

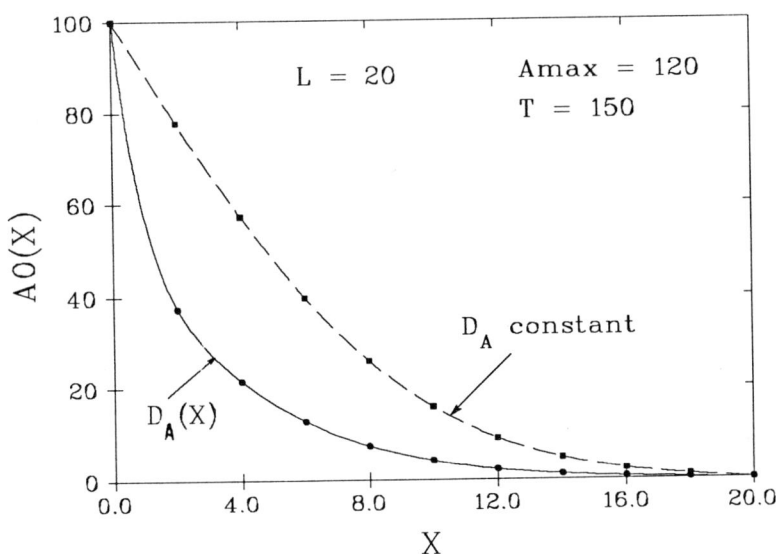

Fig. 2.2.3. Effect of a concentration-dependent diffusion constant on the concentration profile, formulated in accordance with equ. 1.4.13. $L=20$.

When non-homogeneity arises (in a structurally homogeneous medium) as a result of a concentration-dependent diffusion constant, then one proceeds via equation 1.4.13. For $A(0) = 100$, $A_{max} = 120$ and $T = 150$ the effect is shown in Fig. 2.2.3, in comparison with the contour for a constant D_A. Of course, no discontinuity is then involved, but the contour is profoundly changed. As one would expect, the steepest gradients are close to the [A] reservoir.

2.3 Diffusion in the Presence of Electric Fields.

For the exploration of electric field effects, equation 1.4.10 is immediately convenient. Of course, the effects are polarity dependent. Positive fields tend to fill the diffusion medium with [A], negative fields tend to keep it empty. Certainly, enormous changes can be brought about by superimposing fields upon the diffusion process. When $D_A=1/6$ and $E>=3$, the concentration of [A] is almost uniform near the [A]-reservoir, meaning that the transport mechanism is primarily a field current in that region; see Fig. 2.3.1. Towards the right, where the concentrations are kept low by the presence of an "infinite sink" at $X=L$, the transport mechanism is primarily a diffusion current. When $E = 6$, the medium soon becomes completely filled; when $E = -6$ it becomes

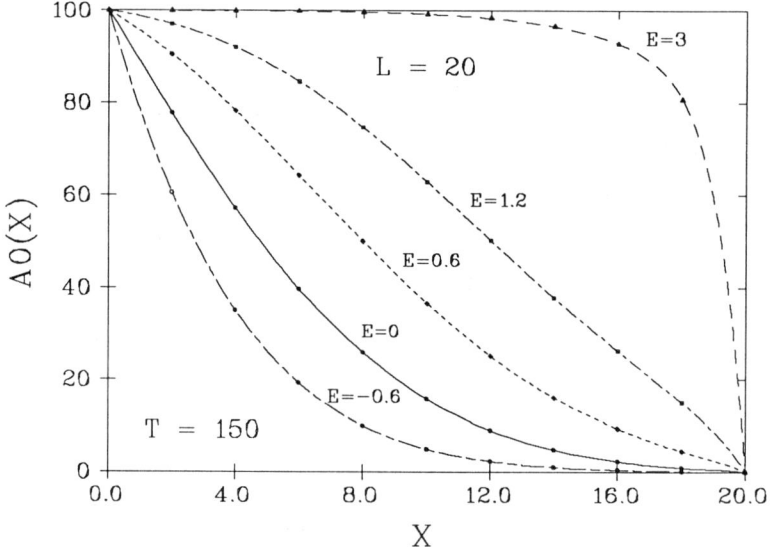

Fig. 2.3.1. Effect of electric fields on concentration contours, computed in accordance with equation 1.4.10. $L=20$, $D_A=1/6$, $T=150$.

empty. This removes all temptation to explore E-values outside that range. In any event, and for reasons which are not yet mathematically clear, $-1<D_A.E<1$ seem to be the stability limits. When E is appreciably outside this range, the computer shows its displeasure by returning alternating positive and negative concentrations! Indeed, it is prudent not to approach too closely to this territory. See also Section 5.8).

2.4 Double Diffusion; Concentration Products.

For many (but not all) of the discussions in later chapters, great importance will have to be attached to concentration products. In particular, we will be concerned with two reagents [A] and [B], diffusing into the medium (of length L) from opposite sides. A value of $A(X).B(X)$ will evidently be associated with every location X, and for the arguments on nucleation which follow, that product will always have to be equal to or greater than a specific level, called K_{SP} (the "precipitation product"). See Section 3.1. The calculation of this product does not involve any new computational procedure; we merely do to $B(X)$ what we are already accustomed to doing to $A(X)$, and then multiply corresponding values. When an electric field is present, we use equation 1.4.10 for the transformation of the A-array. For the [B] reagent, one must (in the present context) expect ionic charges of opposite sign, and allowance for this fact can be made by changing E to $-E$. A positive field thus tends to move [A] towards the right (greater values of X), as already shown, and [B] towards the left.

Once the above transformation has been iteratively computed for $A(X)$ and $B(X)$, four questions are of interest:

(a) Where does the maximum of $A(X).B(X)$ occur in the simplest case?
(b) How does the position of that maximum depend on the boundary concentrations?
(c) How does it depend on the ratio of diffusion constants?
(d) How does it depend on the presence of an electric field?

Questions (a) and (b) are answered by the computed results in Fig. 2.4.1a. When $A(0)=B(L)$, we fully expect to find the concentration product maximum in the middle (at $L/2$) and there, indeed, it is. When the boundary concentrations are *unequal*, one's intuition (a notoriously unreliable tool) suggests that the maximum should be shifted, but it is not. As long as the system is very simple. This matter can also be proved analytically. In contrast, Fig. 2.4.1b shows that the ratio of diffusion constants does indeed affect the position of the maximum, at any rate during the early stages. It shifts it in the direction one would intuitively expect (alas, only a transient reprieve for that faculty); see Salvinien and Moreau (1960) and Lendvay (1964). The peak increases very slowly while the A and B values are low, then very rapidly. However, it must ultimately reach a constant value of $(A(0)/2)*(B(L)/2)$ in the middle, when both concentration profiles are linear, irrspective of D_A and D_B.

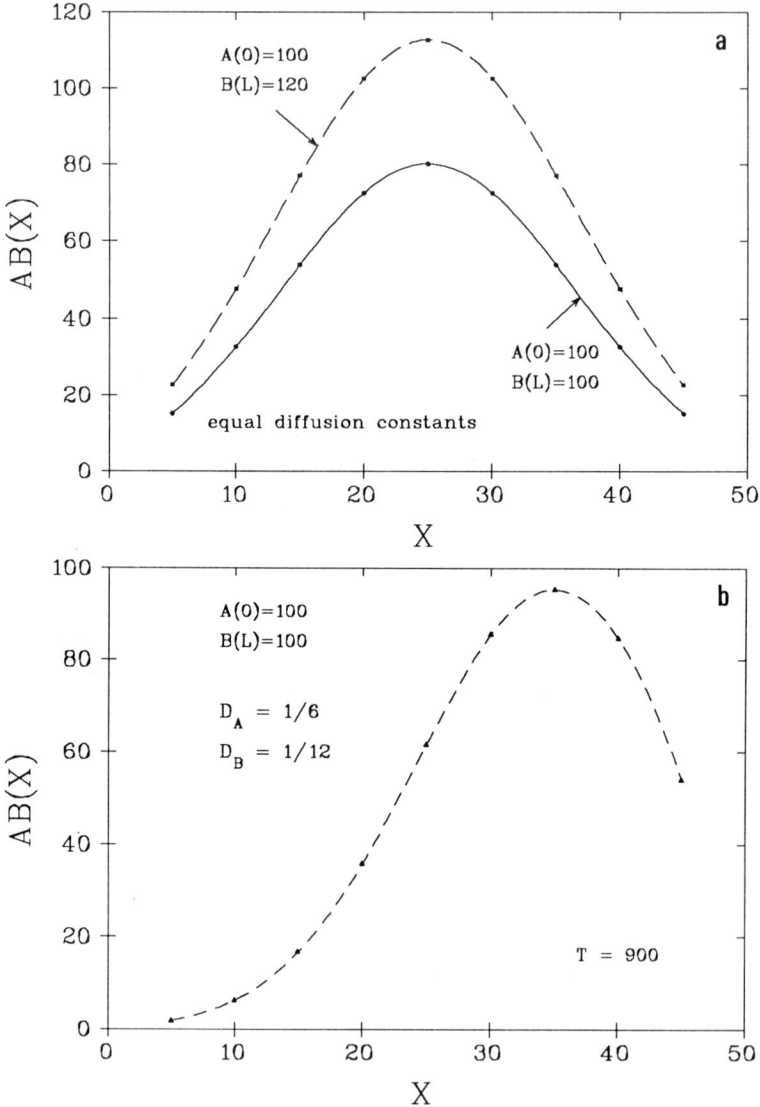

Fig. 2.4.1. Double diffusion; position of the concentration product A.B maximum, (a) for different boundary concentrations and identical diffusion constants (=1/6), (b) for identical boundary concentrations and different diffusion constants.

The results of Section 2.3 suggest that electric fields should similarly shift the position of the concentration product maximum, and in principle they do. However, when $D_A = D_B$ and the reservoir concentrations are the same for [A] and [B], then everything is symmetrical and no field effect is expected. It can be shown that this is still true even when the reservoir concentrations are unequal; the concentration product maximum remains in the middle of the diffusion space, even in the presence of a field. Fig. 2.4.2 explores situations which arise when the diffusion constants (governed by b_A and b_B) are *unequal*. Two things can be seen to happen: (1) the maximum is evidently displaced (in the expected direction) by the field, and (2) the amount of displacement is a function of time.

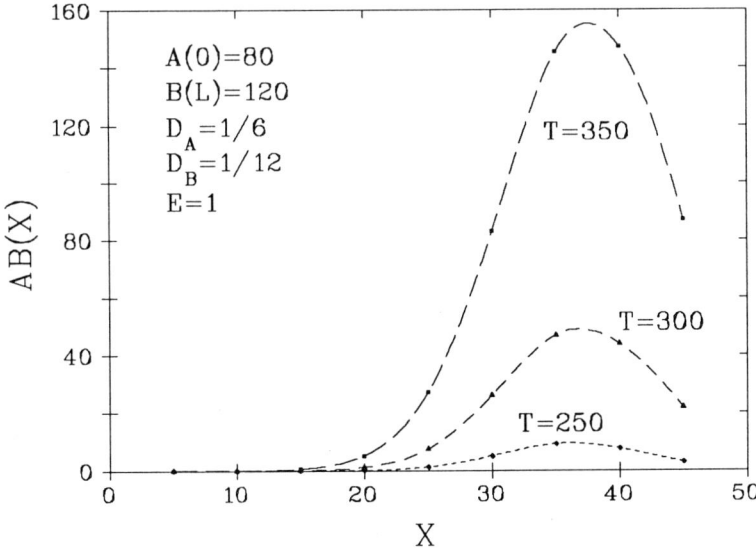

Fig. 2.4.2. Effect of an electric field on the concentration product maximum; unequal diffusion constants: $D_A=1/6$, $D_B=1/12$.

3. NUCLEATION ALGORITMS

3.1 Solubility Relationships in Binary Systems.

The subject of nucleation is highly complex, and its subtleties are outside the scope of this text. For a brief overview of the manner in which the concepts affect our view of precipitation, see Henisch (1988); for more basic treatments, see Mullin (1961), Van Hook (1963), and Zettlemoyer (1969). For present purposes, it will be sufficient to accept the principal findings, and implement them in programing terms. It will be assumed that two diffusing reagents [A] and [B], typically ions in a solution, come together to form a sparingly soluble compound [AB], which constitutes the precipitate. (In Chapter 6 we shall also consider the precipitation of monomers). Whenever [A], [B] and [AB] co-exist in equilibrium, we must have $A(X).B(X) = K_S$, where K_S is called the *solubility product*. This fact follows quite simply from the Law of Mass Action (and no one would be foolhardy enough to challenge that!), but it tells us nothing about how fast or slowly that equilibrium will be approached. As long as $A(X).B(X) < K_S$, for any reason, more [AB] will dissolve. If, on the other hand, $A(X).B(X)$ were greater than K_S (again, for any reason), material would "condense" on any [AB] material already existing, and augment its size. "Already existing" are here the operative words; the formation of new [AB] is something else (see below). When the concentration product equals K_S, the solution is said to be *saturated*. It is said to be *supersaturated* when the solute content is greater than that, and the degree of supersaturation is defined by $(A.B)/K_S$.

When there is no pre-existing [AB] deposit, it is generally found that the concentration product has to be greater than K_S to create one spontaneously, say $A(X).B(X) >= K_{SP}$, where K_{SP} is called the *precipitation product*. Thus $K_{SP} > K_S$, whereby the ratio of the two quantities may be small or large. By way of example, van Hook (1938) gives $K_{SP}/K_S = 5$ for silver chromate solutions, but the ratio for many substances is much higher. Again, the *speed* of precipitation when $A(X).B(X) >= K_{SP}$ is open to discussion. We shall see in Section 3.2 that it depends, amongst other things, on stoichiometric considerations.

At this stage it is necessary to make the familiar distinction between homogeneous and heterogeneous nucleation. Heterogeneous nuclei are foreign substance (whether amorphous particles or crystallites) on which [AB] can be deposited with relative ease, often (in the case of crystallites) epitaxially. When such nuclei are present, K_{SP} may be only slightly greater than K_S for deposition to take place.

If there were no heterogeneous nuclei, or if their number had already been exhausted, the possibility of forming homogeneous nuclei (socalled because they consist of [AB] itself) would remain. Just about all theories of homogeneous nucleation are based on the "critical nucleus", and Ostwald (1897, a and b) demonstrated the physical reality of that concept. [A] and [B] particles in solution form a (supposedly) spherical [AB]-matter, and it can easily be shown that as long as this agglomerate is very small (below a "critical size"), it is unstable. It is stable only when a critical size is reached or exceeded. The formation of such a nucleus demands a free energy increment W_C, and the probability of achieving such an increment should be temperature-dependent in accordance with $\exp(-W_C/kt)$, where k is Boltzmann's constant and t the temperature. There are also sophisticated calculations (based on the statistics and dynamics of the system), aimed at predicting how that probability should depend on the supersaturation, e.g. see Zettlemoyer (1969). The conclusion is that the nucleation probability is *very* low, as long as the supersaturation is low, but it increases sharply when a certain minimum supersaturation is exceeded. That minimum can be readily identified with the condition $A(X).B(X)=K_{SP}$. Nucleation *can* occur at lower concentration products but is then very unlikely. This reasoning also explains why supersaturated solutions can sometimes exist for a long time, even though they are not in equilibrium.

While $A(X).B(X)>=K_{SP}$ is thus a necessary condition for nucleation, we will show below that it is not actually a sufficient one. The term "precipitation" is deemed to refer to the composite phenomenon of nucleation and subsequent growth. *Growth* can take place at concentrations lower than those needed for nucleation, as long as $A(X).B(X)>K_S$. We will assume for the sake of simplicity that when nucleation takes place, identical nuclei of critical size will be formed at any location X. Their number would, of course, depend on the size of the critical nucleus and the degree of supersaturation.

3.2 Stoichiometric Considerations.

We saw in Section 2.4 that when [A] and [B] diffuse into the medium from opposite directions with equal diffusion constants, the concentration product maximum will always be in the middle, irrespective of the boundary concentrations. At some stage during the proceedings, that maximum may come to exceed K_{SP}, and if what we have heard so far were the whole truth, the first deposit would always be found at that place ($X=L/2$). In practice, this is not found to be so, and this means that $A(X).B(X)>=K_{SP}$ is not a sufficient condition, e.g. see Kirov (1969, 1972), and Garcia-Ruiz and Miguez (1982). The complicating factor, not touched by the above arguments, is the matter of stoichiometry.

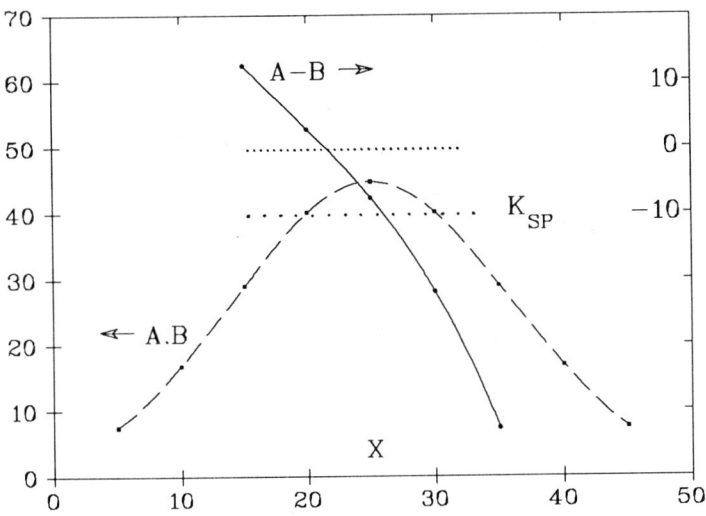

Fig. 3.2.1. The nucleation conditions; stoichiometric and K_{SP} considerations. Example for $L=50$, $K_{SP}=40$, $A(0)=50$, $B(L)=150$, $D_A=D_B=1/6$, $T=600$.

The condition $A(X).B(X)=K_{SP}$ defines a rectangular hyperbola, along which $A(X)/B(X)$ varies enormously, whereas the formation of the [AB] species demands something close to unity for that ratio. When $A(X)/B(X)$ is either very high or very low, the formation of [AB] is very unlikely; see Henisch and García-Ruiz (1986). For such a binary compound, the near-equality of $A(X)$ and $B(X)$ thus becomes a second necessary condition of nucleation. Considering (at some time T) all the continuum of X over which $A(X).B(X) >= K_{SP}$, nucleation will occur first at the particular value of X for which the near-equality condition is also satisfied. When $A(X).B(X)$ first comes to equal K_{SP}, there may be no such point; when the concentration product exceeds K_{SP} there may actually be many such points which fail to qualify on stoichiometric grounds, but in due course, one point *will* qualify. Fig. 3.2.1 illustrates some of the possibilities for $A(0)=50$ and $B(L)=150$. The concentration product maximum is in the middle, as expected. The full line displays $A-B$; note the scale on the right. If K_{SP} were 40, as suggested by the horizontal dotted line, then $A=B$, which occurs at $X=21$, would be just inside the precipitation range. If, on the other hand, K_{SP} were set (say) 43, then the equality point would be outside the range, and no precipitation would occur (or, at any rate, not at $T=600$).

The question is how the "*near*-equality" is to be defined, and the answer is obvious: in the first instance: by reference to an appropriate index R_R, defined as:

$$R_R = \text{Abs}[A(X)-B(X)]/[A(X)+B(X)] \qquad (3.2.1)$$

After every pass (from $X=1$ to $L-1$) aimed at re-calculating $A(X)$ and $B(X)$, it would be necessary to ask two questions (again for every value of X): (1) is $A(X).B(X) >= K_{SP}$, and (2) is R_R small enough (and how small is that)? We could employ a yes/no test, but the nucleation probability N_P is actually expected to be a *continuous* function of R_R, and the form of this function is known to be roughly Gaussian; see Henisch and García-Ruiz (1986). Without any ceremony, we therefore define N_P as:

$$N_P = \exp[-(R_R/E_R)^2] \qquad (3.2.2)$$

where E_R is an adjustable parameter, called the "equality range". It governs the "fatness" of the Gaussian curve. Assuming $A(X).B(X) >= K_{SP}$, nucleation is, of course, a certainty where $R_R=0$, no matter what the value of E_R may be. Setting $E_R=1$ makes the bell-shaped probability curve exceedingly wide and, for all practical purposes, turns the stoichiometric requirements *off* (compare Section 5.7). This is sometimes useful for visual illustration purposes, because nucleation then occurs simply (and very predictably) when $A(X).B(X)=K_{SP}$. On the other hand, $E_R<1$ implies a more realistic model.

One more complication: If equation 3.2.2 were in sole charge then, no matter what the value of R_R, there would always be some probability of nucleation. On the other hand, we cannot have "nucleation" when there is less than one grain nucleating; there has to be a cut-off somewhere. We propose to make that cut-off here schematically by adding a further clause to the program. In Basic:

```
if NP<0.2 then let NP=0
```

The continuity of equation 3.2.2 is thereby maintained over its useful range. Of course, the 0.2 is merely a plausible choice for the cut-off; other choices could be made.

Once $N_P(X)$ has been established by means of equation 3.2.2, we have the basis for two further calculations: (a) How many grains are formed? (b) How should the concentrations $A(X)$ and $B(X)$ in solution be decremented to maintain the xero-sum character of the transaction? After all, if the material is in the newly formed nucleus, it cannot be in the solution at that time; see Section 3.3.

3.3 Graincount and Grain Size; Concentration Decrements.

Each grain formed (initially, each critical nucleus formed) will consist of a certain mass, say M_{PG0} (for "mass per grain at time zero"). In this total, particles of [A] and [B] will be counted separately, even though their numbers must, of course, be equal. If $N(X)$ such nuclei are formed at X, then $N(X).M_{PG0}$ must be the amount of material withdrawn from the solution at that place. M_{PG0} is evidently a measure of the size of the critical nucleus but, with great regret, we do not generally know what that size is. There is, however, a way of converting this ignorance into an advantage or, at any rate, into a desirable flexibility: we can make M_{PG0} an adjustable parameter, and bask in virtue thereafter. On that basis, we could then calculate the graincount $N(X)$, if we knew the amount of solute by which the solutions have to be decremented. It turns out to be simplest to calculate that decrement first.

Once the nuclei are formed, they co-exist with solution. One might assume that this equilibrium is instantly established in a single step at the moment of precipitation (e.g. on the grounds that we have only a very small interval on either side of X under consideration, but it would be prudent to allow for other possibilities (see below and Section 4.1). Even so, that simple assumption will here be made. In as much as equilibrium, $A(X).B(X)=K_S$, defines a rectangular hyperbola, the solute content of the solution will have to be decremented from the original $A(X)$, $B(X)$

to a point $A(X)$-D_{EC}, $B(X)$-D_{EC}, which must be on the rectangular (K_S) hyperbola; see Fig. 3.3.1. For the binary compound here envisaged, the decrements are, of course, the same for [A] and [B]. This means that

$$[A(X)\text{-}D_{EC}].[B(X)\text{-}D_{EC}] = K_S \tag{3.3.1}$$

has to be solved for D_{EC}, and that can always be done. No need for numerical methods, it is done analytically. The resulting D_{EC} is specific to the location X under review. In that sense, it should perhaps be written as $D_{EC}(X)$, but in programming terms this is not necessary; the D_{EC} values are not needed for later purposes, and need not, therefore, be retained in any dimensioned array.

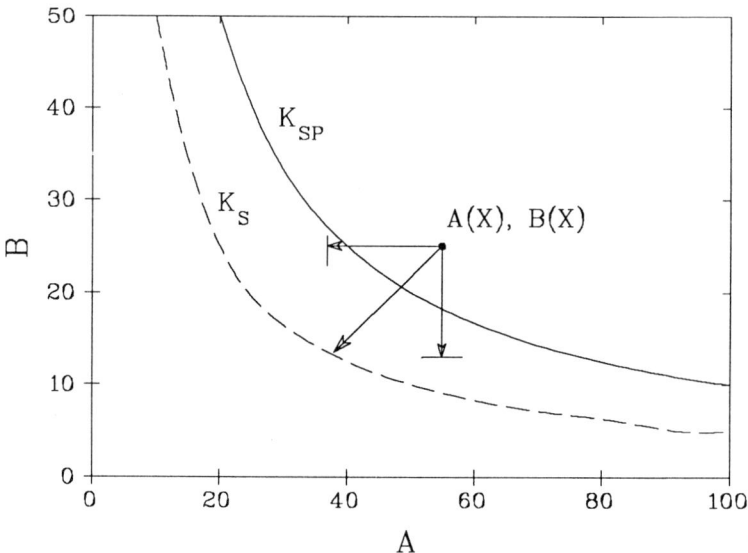

Fig. 3.3.1. Decrementing the solute concentration at nucleation. Points $A(X)$ and $B(X)$ are assumed to satisfy the nucleation conditions. Long arrow shows direction of change, short arrows show actual decrements for $N_P=1$.

One more refinement: that full decrement would be appropriate if nucleation at that point were a certainty. Since it is only a probability, the appropriate decrement is its "probability value", that is to say, $N_P.D_{EC}$. Though D_{EC} is calculated by means of equation 3.3.1, we subsequently adjust the two concentrations (in terms of Basic) in accordance with:

```
let A(X) = A(X) - NP*DEC
let B(X) = B(X) - NP*DEC
      (new)      (old)
```

This is done in the course of a pass from $X=1$ to $L-1$, but only when N_P is non-zero will an actual concentration adjustment be made. Thus we can proceed to calculate $N(X)$ by putting (in Basic)

```
if N(X)=0 then let N(X) = 2*NP*DEC/MPG0
```

where the factor 2 comes from the fact that each constituent is counted separately. The conditional clause makes sure that nucleation occurs only once at each site, unless for some reason or other that site has been emptied by re-solution. M_{PG0} is here expressed as an equivalent concentration. Thus, $M_{PG0}=0.1$ is a plausible starting point for explorations, but is not intended to mean that the mass of the critical nucleus is 0.1 kilograms or 0.1 particles! It means that the critical nucleus contains the same mass as would be contained in unit volume of a solution of (normalized) concentration 0.1. (Compare the reservoir concentrations here used, typically 100. In that sense, $M_{PG0}=0.1$ is actually high.)

Indoctrination interlude. The implied normalizations here used represent an enormous convenience, but also call for a flexible attitude to the actual numbers involved. Thus, we know very well that $T=20$ stands for a time that is twice as long as $T=10$, but does not stand for 20 seconds. T is a "computer time". Other numbers are similarly relative, and serve flawlessly in that capacity for most purposes. Their translation into absolute real-world terms is another matter, always possible in principle, but frequently cumbersome. Indeed, in most practical cases, we do not *have* known values for some of the essential parameters which enter into the processes. The size of the critical nucleus is is merely one such example. In absolute terms, this lack of data would bring the exploration to a standstill; in relative terms it need not do so at all. The purpose of the exercise is in any event to study, clarify and predict modes of behavior, and not (or, at any rate, not necessarily) to calculate specific features in terms of laboratory magnitudes.

We have here assumed that K_S is an immutable constant (governing the solubility), but the solubility of [AB] could in principle be influenced by other substances that might be present or, for instance, by the varying pH of the solution. In Section 5.6) the constancy of K_S will be relaxed, in order to leave room for interactions between [AB] and secondary reaction products. For example, the primary reaction between [A] and [B] may lead to changes of pH, and those, in turn, may affect the solubility of [AB].

4. PRECIPITATE GROWTH AND RE-SOLUTION

The considerations of Section 3.3 left us with $N(X)$ assemblies of [AB] nuclei of critical size at points X, in equilibrium with the surrounding solution, now appropriately depleted of solute content. Accordingly, local concentration gradients arising from the precipitation are now superimposed on the overall pattern. As the diffusion processes continue, $A(X)$ and $B(X)$ will, of course, change, and two sets of questions arise:

(a) how will the *changed* values of $A(X)$ and $B(X)$ interact with the deposit, and
(b) how will the deposit affect the chances of nucleation at neighboring sites.

Question (a) concerns precipitate growth, precipitate re-solution and the corresponding adjustments of solute concentration (Section 4.1). Question (b) addresses the origins and development of periodic structures (Section 4.2).

4.1 Precipitate - Solution Interaction; Rates of Mass Transfer.

Beginning with the (just) decremented values of $A(X)$ and $B(X)$, there are evidently two possibilities: these concentrations may increase or decrease as a result of further diffusion. In the presence of a deposit, it is highly unlikely that the concentration product will ever reach K_{SP} again, but if that product came to be higher than K_S, the deposit (originally at equilibrium) would grow; if it were to fall below K_S, it would begin to re-dissolve. In each case, the computer implementation of the changes is similar to that discussed in Section 3.3. There is, however, one change. In Section 3.3, we adjusted the concentrations by the *probability value* of the calculated decrement, but here we do not have a multiplier of the same kind. Yet, one does want to make allowance for the fact that growth may depend not only on the concentration product, but also on microscopic processes (not here under investigation) at the growth surface. For that purpose we introduce two adjustable coefficients, R_{SC}, the *re-solution coefficient*, and G_{RC}, the *growth coefficient.* These coefficients are then used just as N_P is used above; likewise, they are equal to or smaller than unity. We shall see below how one might arrive at plausible values of G_{RC} and R_{SC} for use in the computations.

The first step of the implementation is to calculate the decrement. For the case of growth, this is now called D_{ECG}, to distinguish it from the original D_{EC}, above. This is again done analytically by means of an equation of the form 3.3.1. Once D_{ECG} is given, we proceed to decrement the solute concentrations in accordance with (in Basic):

```
let   A(X) = A(X) - GRC*DECG
let   B(X) = B(X) - GRC*DECG
```
(4.1.1)

and then augment the grain size by the corresponding quantity. In general, grain size is measured by $M_{PG}(X)$, which has a minimum value M_{PG0}, as we have seen. Thus, we have here for the *first* pass after nucleation:

```
let  MPG(X) = MPG0 + 2*GRC*DECG/N(X)
```
(4.1.2)

since the total amount of available material must be shared between the $N(X)$ grains already formed. [Note that steps must evidently be taken to prevent the calculation of $M_{PG}(X)$ when $N(X)=0$!] For later passes, the equation will, of course be:

```
let MPG(X) = MPG(X) + 2*GRC*DECG/N(X)
    (new)    (old)
```
(4.1.3)

The corresponding code lines for the case of re-solution are:

```
let A(X) = A(X)+RSC*INCS
let B(X) = B(X)+RSC*INCS
```
(4.1.4)
```
let MPG(X) = MPG(X)-2*RSC*INCS/N(X)
    (new)    (old)
```
(4.1.5)

where we use I_{NCS} (INCrement corresponding to Solution) in place of D_{ECG}, since the reagent concentrations are actually augmented by the change. The value of I_{NCS} is obtained as the appropriate solution of

$$[A(X) + I_{NCS}].[B(X) + I_{NCS}] = K_S. \quad (4.1.6)$$

Compare equation 3.3.1.

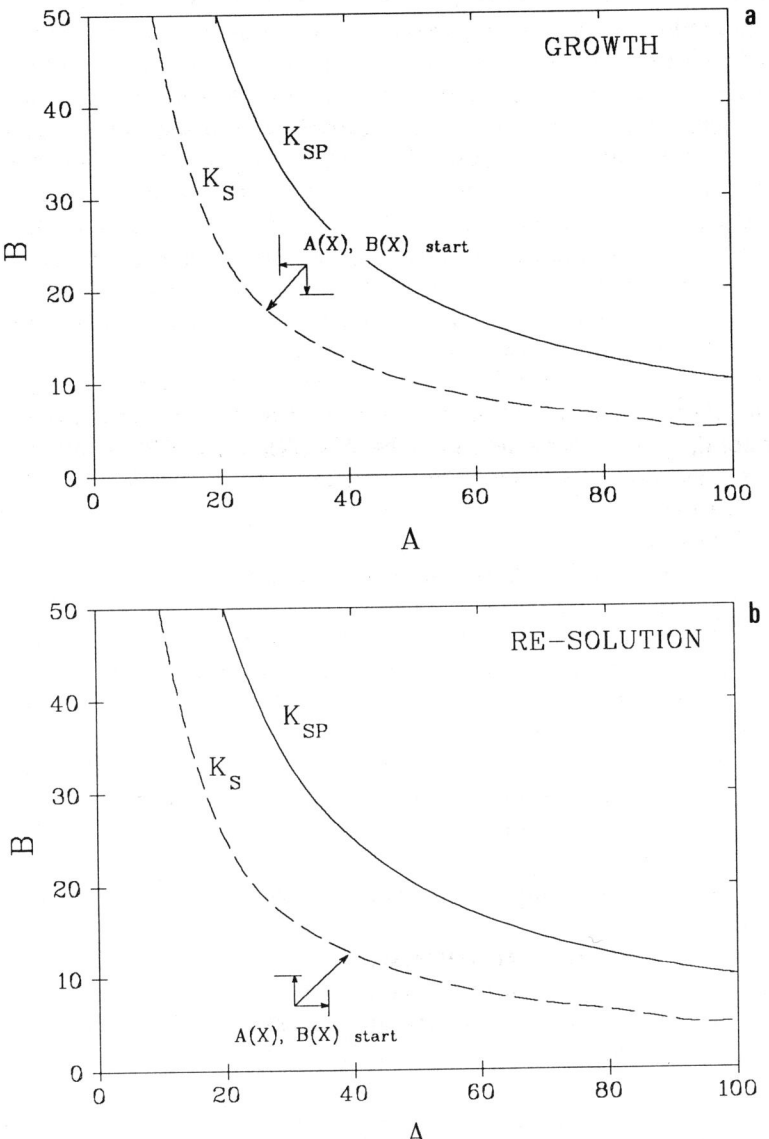

Fig. 4.1.1. Mass transfer resulting from a non-equilibrium value of the concentration product at a location X.
(a) $A(X).B(X) > K_S$; deposit growth, reagent concentrations decremented ($R_{SC}=0.5$), (b) $A(X).B(X) < K_S$; re-solution, reagent concentrations incremented ($G_{RC}=0.5$).

Fig. 4.1.1 illustrates growth and re-solution. Fig. 4.1.1a, for growth, is actually similar to Fig.3.3.1, except that the starting point is now between the K_S and K_{SP} hyperbolas.

Depending on the values of R_{SC} and G_{RC}, the ongoing diffusion processes may lead to higher or lower concentrations $A(X)$ and $B(X)$ during subsequent passes, and hence to different concentration products. The same considerations will then have to be applied again.

There remains the problem of arriving at plausible values of R_{SC} and G_{RC}. In doing so, we have an opportunity of making allowance for some well-known facts: (a) small grains dissolve more easily than large ones, and (b) large grains grow more rapidly than small ones. Indeed, they tend to grow at the expense of small ones; see Chapter 6. So, one might put (in Basic):

$$\text{let RSC = [MPG0/MPG(X)]}^\wedge(1/3) \tag{4.1.7}$$

which obviously begins by being 1 and diminishes as the grains grow. The (1/3) power ensures that it does so as a function of the radius, maintaining the useful fiction that the grains are spherical.

For G_{RC} we must make a different kind of provision; the coefficient should *grow* (though not beyond unity) in proportion to the grain surface area. This is ensured by an expression of the type (in Basic):

$$\text{let GRC = 1 - 0.9*[MPG0/MPG(X)]}^\wedge(2/3) \tag{4.1.8}$$

which begins by being very small (0.1), and then grows to unity as the grains acquire mass. The 0.9 factor is actually arbitrary, and no particular claims of scientific profundity are made on its behalf. It ensures, though, that all the grains can grow, whereas a factor of 1 would have prevented any growth from even beginning, since $M_{PG}(X)$ must equal M_{PG0} before it equals anything else. For a very different approach to the problems of particle size, growth and re-solution, see Sections 6.2 - 6.4.

4.2 Development of Periodic Structures.

This section (and the next) are not really about further programing, but rather about the consequences of the programing provisions already made. An additional (but, of course, purely optional) feature might be one that identifies and records the times of nucleation, i.e. the times T at which M_{PG0} turns to $M_{PG}(X)$, noting that this can occur only once for each location. The values of T so obtained might be simply displayed on the monitor, or else stored in a special array. They could then be used for the timing exploration described below.

To return to the arguments of Section 4.1 in somewhat greater detail, let us consider a location X (e.g. $X=44$ in Fig. 4.2.1) at which the concentration product had been rising for a while until the conditions of nucleation were satisfied. As a result of that nucleation, the concentrations A and B were decremented. If N_P had been equal to unity in this case, the decrements (equal for [A] and [B]) would have brought the concentrations to their equilibrium values, for which $A(X).B(X)=K_S$. For $N_P<1$ the excursion would be in the same direction, of course, without actually reaching the equilibrium level. After nucleation, the location X is in any event a solute sink. Growth of the nuclei at that point will always tend to diminish the local concentrations, and diffusion (to the sink) from adjoining cells may or may not manage to cope with this need for material. If it *just* does, an unlikely situation, a steady state is reached. If the influx is insufficient, the concentration products will further diminish, first to K_S and later, conceivably, even below it. This tends to be a self-rectifying process, since $A(X).B(X)<K_S$ implies re-solution and, thereby, the *release* of solute into the medium. Because diffusion must necessarily concern more than a single cell, an effective sink will lower the reagent concentrations not only at X, but in several cells on either side of X. In those cells there is then a greatly diminished likelihood of satisfying the precipitation conditions; these regions are "starved of nutrient". They will tend to remain free of deposits. However, at some distance from the original nucleation site, the conditions *can* once again be satisfied, and a new deposit will result where and when they are. This establishes the striated pattern of deposits widely known as *Liesegang Rings*; see Figs. 1.1.1 to 1.1.3.

Whether such a pattern of deposits ever reaches stability depends on many parameters of the system, notably including the initial reservoir concentrations and the extent (if any) to which they become depleted by the diffusion process; see Sections 4.6 and 5.3. In general, a system may change slowly or fast in the course of time, with each deposit pursuing its own career. Deposits, once formed, may re-dissolve, and disappear altogether. Alternatively, a thick deposit may re-dissolve on one side and grow on the other, thereby appearing to "move". Movement of the kind here envisaged has often be observed, e.g. in the study of urea deposits by Knöll (1938 a and b); see also Section 6.2. For all these reasons, the term "periodic" used for such precipitate patterns must be interpreted with liberal elasticity. Deposits may be widely or closely spaced, and are occasionally so close as to be "unresolved"; the deposit is then for all practical purposes continuous; Deposits may be thick or thin (see Section 4.4). In general, their detailed behavior depends on the system parameters in a manner too complex to be intuitively predictable, and likewise too complex for analytic treatment. A computer analysis is needed precisely for this reason.

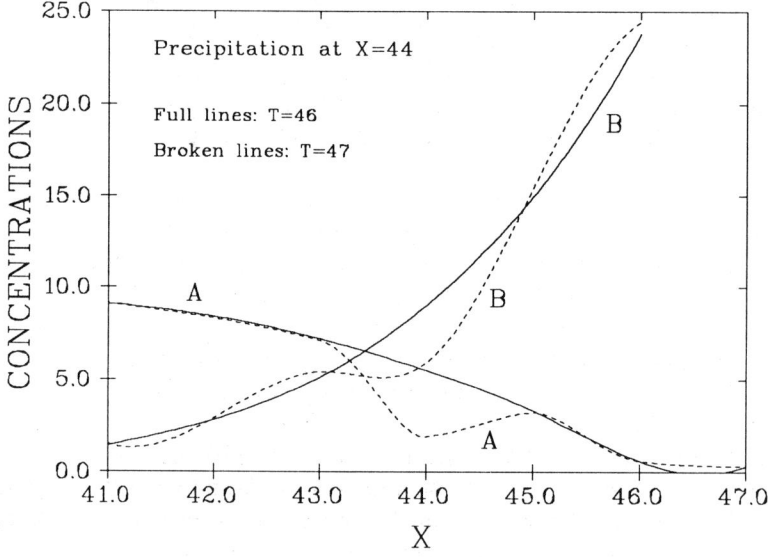

Fig. 4.2.1. Concentration profiles during the formation of a precipitate at $X=44$; Example: diffusion medium pre-charged with reagent [A] to a concentration 10. Reagent [B] diffusing from the right. $L=50$, $B(L)=100$, $K_{SP}=50$, $K_S=10$, $D_A=B_B=1/6$.

Analogous arguments can also be applied to biological systems. Thus, Knöll (1939) has demonstrated periodic structures consisting of *bacterium vulgare (proteus)* colonies. Although, for the same of an example, the present discussion is formulated in terms of a binary ionic system, the phenomena, in the laboratory and in nature, are far more general than that; see also Borchert (1989) and Chapter 6.

In passing one might note that the exact curvatures shown in Fig. 4.2.1 are conjectural; we have computed data only for integral values of X. This is not merely a point about graphics presentation, but has more profound implications; see Section 4.4. Note also that before the nucleation event at $X=44$, all [A] diffusion is to the right, all [B] diffusion to the left. After nucleation, the pattern is much more complex.

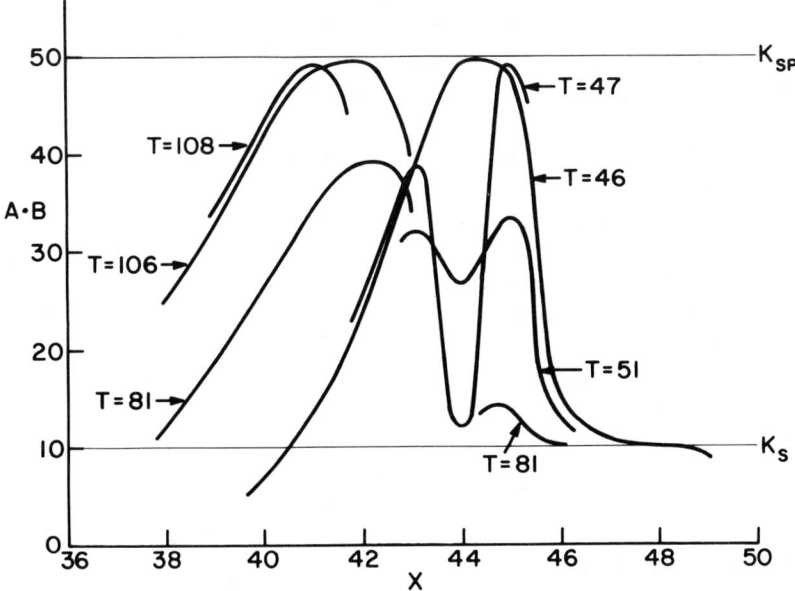

Fig. 4.2.2. Concentration product profiles during the formation of precipitates at $X=44$ and $X=42$. Example: diffusion medium pre-charged with reagent [A] to a concentration 10. Reagent [B] diffusing from the right. $L=50$, $B(L)=100$, $K_{SP}=50$, $K_S=10$, $D_A=D_B=1/6$.

Fig. 4.2.2 gives corresponding concentration product contours in successive stages from $T=46$ to $T=108$. The ongoing diffusion causes $A(44).B(44)$ to rise after $T=47$, and similarly causes the peak at $X=43$ to be depleted. The curve for $T=51$ shows this. In this particular system, at the times discussed, deposits are already in existence at $X=46$, 47, 48 and 49. In that region, precipitate and solute co-exist, which is why the concentration product approximates closely to K_S. As the peak at 43 diminishes, the new deposit at $X=44$ will soon become part of that extended and growing "equilibrium zone". The $T=81$ contour illustrates this tendency.

At $X=50$ there is, of course, a (here) infinite sink for [A]. Accordingly, the product $A(49).B(49)$ is kept low and, indeed, comes to be below K_S, as shown. The deposit at $X=49$ is therefore in the process of re-solution. (On a graphic display, it is convenient to signify this by assigning a distinctive color this deposit and, indeed, to any other re-dissolving deposit.) In the situation here shown, only one location is so affected, but adjoining cells will be affected in due course. Indeed, the entire (think) first deposit is destined to be wiped out.

The development of the next deposit (at $X=42$) is also shown in Fig. 4.2.2. At $T=106$, the K_{SP} level is almost reached; nucleation happens at $T=107$, and the result is to make the $X=42$ location into a sink ($T=108$).

4.3 Spacing Relationships.

As the processes envisaged above continue, there arises, as we have seen in Fig.4.2.2, a new deposit at $X=42$, but none at $X=43$. By the time the new deposit at 42 is formed, the earlier one at 44 has grown from an initial mass, $M_{PG0}.N(44) = 6.84$ to $M_{PG}(44).N(44) = 53.2$.

As it happens, the next deposit in this particular system (here used for illustrative purposes) forms at $X=39$ ($T=188$). There is never a suggestion that the spacings must be equal, unless some very special measures are taken to make them so. This can be done, over limited distances, albeit only by trial and error choice of the operating parameters. Ordinarily, spacing depends on complex (here deterministic) interactions between the various parameters of the system, and while the series of positions can always be expressed as an empirical "law" as, for instance, shown by Matalon and Packter (1955), Mathur (1961) and Shinohara (1970, 1974), there are no simple interpretations for such relationships. The matter is further complicated by two facts: (a) that it is not always possible to say unambiguously *where* a deposit is, considering that some deposits have complex density profiles, and (b) that deposits can actually "move", as already described above.

By now it should be abundantly obvious that attempts to interpret empirical "spacing laws" are doomed to failure. See Morse and Pierce

(1903) and Wagner (1950) for two such attempts. There is, however, one relationship that appears to be generally maintained, at any rate in simple cases. Equation 1.4.5b encourages a cautious expectation that, for a fixed diffusion constant, events in very simple systems should generally happen at distances X that are proportional to the square root of the corresponding times. Henisch (1988) has already given computed results which show this to be true for different values of K_{SP}, and Fig. 4.3.1 provides an example for two values of $D_A(=D_B)$. The relationship is, indeed, linear for each value of the diffusion constant.

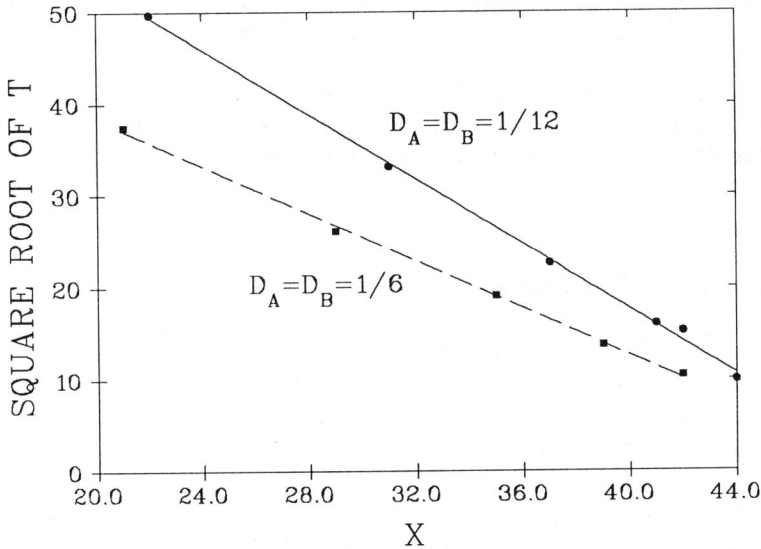

Fig. 4.3.1. Relationship between the time of nucleation and the position of the nucleation site. Example for $B_{R0}=100$, medium pre-charged with [A] to a concentration 10, $E_R=1$. Two values of the diffusion constants. Note that deposit formation begins at high X-values.

Superficially, everything looks as if it were in good order, but inspection shows that the slopes are not in a 2:1 ratio, as one might intuitively expect them to be. Moreover, the deposits are in different places for the two runs. Has anything gone stray? No. The fact is that the phenomena we are monitoring involve more than just diffusion; they involve diffusion, growth and re-solution. Consider a potential nucleation site, say, in mid-range. In the example shown, the [A] reagent is already available (the medium having been pre-charged with it). Reagent [B] diffuses towards the site from the right, and the rate at which it does so is certainly influenced by its diffusion constant D_B. However, the arrival rate is also influenced by the consumption of [B] involved in the growth of deposits already formed, as well as the release of [B] by the (few) deposits re-dissolving. These processes, in turn, influence the pattern of concentration gradients, and thereby also the position of actual nucleation sites. Thus, a change of diffusion constant (or, as here, of both diffusion constants) affects the total outcome in complex ways, even though the linearity of the relationships in Fig. 4.3.1 remains unaffected. See also Fig. 6.1.1. One could say that there is (surprisingly) an *effective* diffusion constant which depends on the actual diffusion constant and on the pattern of precipitates encountered by the diffusing reagents. Thus, the effect of diminishing the diffusion constants by 50% is less than one might think, precisely because the diffusion constants are not (after precipitation) in sole control of the solute arrival rates at any particular site.

The data in Fig. 4.3.1 were obtained with the E_R parameter set to 1, which means that the stoichiometric requirements for nucleation were for all practical purposes ignored. The altogether remarkable thing is that the relationship can be shown to be similarly linear for (say) $E_R=0.1$, despite the fact that the deposits are then in differnt places. A profound relationship is evidently at work. However, in more complex systems in which different parameters come into play at different times, one would not expect any simple relationship to be maintained. Indeed, the inter-deposit spacings may not vary monotonously, There are many reports in the literature of spacings reaching a maximum and then diminishing again ("revert" patterns; e.g. see Kanniah (1983), Packter (1955), and Flicker and Ross (1974). Needless to say, computer models can reproduce those more complex forms of behavior also.

4.4 Deterministic and Probabilistic Models.

In Section 3.2 we introduced an operational variable N_P, which depended on stoichiometry, but although it was called a "nucleation *probability*", its value was deterministically calculated by equation 3.2.2. Indeed, in all the arguments and algorithms used so far, events happen because they had to. Such a model, with zero degrees of freedom, would be totally appropriate if we could claim to know everything there is to know about the system. In practice, we do not, and it is desirable to make some allowance for the remaining margin of ignorance.

Another problem has already been referred to: we compute concentrations, etc. for certain fixed positions X, and every deposit formed is formed (having no choice) at one of those positions. Though it is possible to have adjoining deposits, most deposits calculated in this way are only one cell thick. They have no density profile. They have a mass, but that mass is, by implication, regarded as uniformly distributed over the cell width. Moreover, *between* deposits, there is supposedly no mass at all. In practice, things are often very different. A Liesegang Ring system is rarely an "all or nothing" affair. The principal deposits *have* a finite thickness, the mass *can* vary over that thickness, and a small amount of material can and does often appear in the spaces between principal deposits. See also Sections 6.2 - 6.4.

The problem is how to reach an accommodation with these facts. It can be done by introducing a random element into the procedures. We already impose two conditions for nucleation, one depending on $A(X).B(X)$, and one depending on $A(X)-B(X)$. Of course, there may be other relevant circumstances; we need not specify what they are, which is just as well, since we do not precisely know them. We need only say that *if* the above conditions are satisfied, nucleation will *probably* happen, not certainly. Of course, once we are in the business of introducing an element of ignorance into the algorithm, we have enormous scope! There are, no doubt, many ways of doing it, of which two will be described. One could, for instance, allow nucleation to happen more or less copiously, depending on a random element (see Method A, below). Alternatively, one could hold that nucleation is an all-or-nothing affair, but that the likelihood of it happening at all involves the random element (see Method B, below).

In either case, nucleation that *could* happen at a particular X because the nucleation conditions are satisfied, would occur only sparingly, or even not at all. As a result, the concentrations will not be decremented less than they would otherwise have been. Normal diffusion will continue, and the concentration product will further increase until nucleation occurs at some other value of X, ordinarily not far away. The pattern of deposits (the Liesegang Ring system) would then be very different. Because each such pattern would be (merely) the result of a series of "controlled accidents", it would be of no special significance. What would be of significance would be the *average* over several runs; see Henisch (1988). Such a series of averages would reflect deposits of greater-than-single cell thickness, with a computed mass profile; see Section 5.9.

48 Periodic Precipitation

Fortunately, programing provision for such a feature is a simple matter, but it consists necessarily of four parts:

(1) *Introduction of the random element.*

As discussed above, there are (at least) two methods of imposing a random character element upon the proceedings. They are described as Method A and Method B below. [In the supporting software available for this book, the choice between Methods A and B is made via the N_{UC} parameter, which is 0 for A and 1 for B.] Fig. 4.4.1 gives a comparison of the corresponding nucleation probability characteristics, both for $D_{NC} = 0.5$. In Method A the probabilities are scattered; in Method B they either have their full value, or else are zero (missing points on the curve).

METHOD A:

It is convenient to introduce the random element by modifying equation 3.2.2. For this purpose, we define a "deterministic nucleation coefficient" D_{NC}, user determined and variable between 0 and 1. This coefficient is then used in conjunction with the output of the random number generator (provided by just about every computer language). In such terms, equation 3.2.2. becomes (in Basic):

```
let  NP = [DNC + (1-DNC)*RND]*exp[-(RR/ER)^2]           (4.4.1)
```

where RND generates a random number between 0 and 1. Test: when $D_{NC}=1$, the random term disappears; the program then runs deterministically. When $D_{NC}=0$, the operation is totally probabilistic. Since D_{NC} can be arranged as a user-input, the degree of probabilistic operation is under user control.

METHOD B:

Here we also use a "deterministic nucleation coefficient", D_{NC}, but in a very different way. In Basic:

```
            if RND<DNC then
                 let Z=1
              else
                 let Z=0
            end if
                                                         (4.4.2)
            let NP = Z*EXP[-(RR/ER)^2]
```

Note that in Method A, D_{NC} can vary freely between 0 and 1. Here $D_{NC}=0$ would lead to zero deposition. That fact should guide the choice of D_{NC} values.

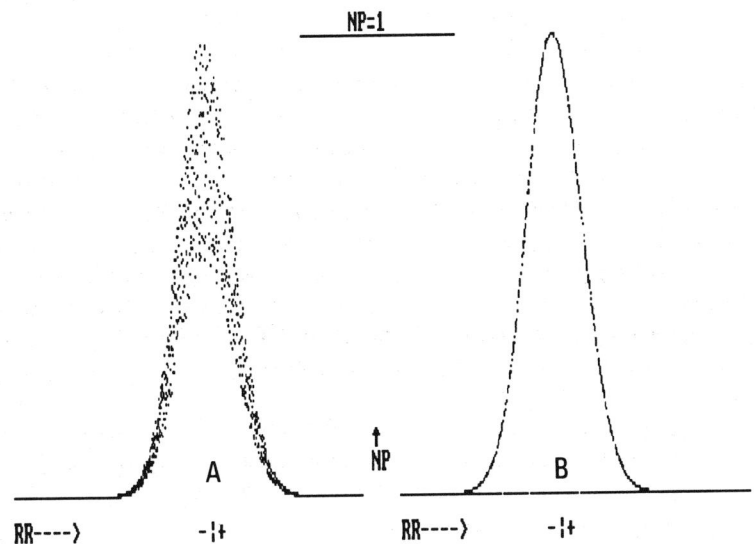

Fig. 4.4.1. Two methods of introducing a random element into the algorithm. $D_{NC}=0.5$. Left: Method A; equation 4.4.1, Right: Method B; equation 4.4.2.

(2) *Provision for multiple runs.*

In Section 1.4, we discussed the nature of each pass (from $X=1$ to $X=L\text{-}1$), for each time (from $T=1$ to T_T). This constitutes a "run". Now all we have to do is to repeat the procedure again and again, until the last of the total number of runs, R_T, has been completed. That is to say, we need to code:

```
For R=1 to RT
For T=1 to TT
For X=1 to L-1
    .
    .              [! No readout needed, since
    .               no specific run is significant.]
```

```
            Next X
            Next T
               .              [! Accumulate all the results.]
            Next R
               .                     [! Average the results.]
            Next R.
```

A cross-provision is easily made, whereby $D_{NC}=1$ is always linked with $R_T=1$, and $D_{NC}<1$ with $R_T>=2$.

(3) *Accumulation of results from repeated runs.*

No problem. We first set up two more arrays, and make provision for a third (see below): $T_N(X)$ for the total number of grains, and $TM(X)$ for the total mass. Then, every time we calculate $N(X)$ and $M_{PG}(X)$, we also write:

```
            let   TN(X) = TN(X) + N(X)
            let   TM(X) = TM(X) + N(X)*MPG(X),
```

and thereby sum the totals.

(4) *Averaging and Display.*

After the final run, we make use of a third auxiliary array, namely $TM_{PG}(X)$, and compute its values by means of one more pass:

```
            for X = 1 to L-1
            let  TMPG(X) = TM(X)/TN(X)
            next X
```

Displaying the results is, of course, very necessary, but how it is done is largely a matter of taste. The accumulation arrays $TN(X)$ and $TM(X)$ can be scanned, and their contents divided by R_T for averaging. $TM_{PG}(X)$ is already an average. Since all this would take place at the very end of the main computation, and since the display is performed only once, its speed is not an important consideration; an analog graphic display is therefore very appropriate. Of course, a hard copy of the numbers is easily printed out.

Making provision for a probabilistic feature will give a more realistic appearance to the computed results, and this should make it somewhat easier to match them to observations on actual (laboratory) systems. It also encourages us to take the deterministic results a little less literally than we might otherwise have been inclined.

4.5 Effect of Secondary Reaction Products.

The entire discussion, so far, has been in terms of two "reagents", [A] and [B], which come together to form [AB] and nothing else or, at any rate, nothing else that matters. This is not necessarily a realistic situation. Take, for example, a typical reaction that is often studied in a diffusion medium, the growth of calcium tartrate crystals. Traditionally, a solution of calcium chloride is diffused into a medium containing tartaric acid. The result is, certainly, calcium tartrate, but also hydrochloric acid, and whereas the tartrate is only sparingly soluble in water, it it appreciably soluble in hydrochloric acid. The production of calciumn tartrate thus comes to a natural end, when the rate of production equals the rate of re-solution. HCl is, in a sense, a "waste product" of the reaction. More generally, one might simply envisage that precipitation of the wanted substance changes the pH of the medium, and that the pH, in turn, has an effect on the solubility product (K_S) of the precipitate, a parameter hitherto regarded as a system-wide constant.

It is a relatively simple matter to make allowance for these possibilities, again to a degree that is under the software-user's control. To do so, we envisage a third reagent, thoughtfully called [H], that is produced locally in proportion to the mass of [AB] precipitated. Concentrations of [H] are then treated in the same way as all the other concentration variables. Thus, (in Basic):

$$\text{let} \quad H0(X) = H0(X) + N(X)*MPG0 \qquad (4.5.1)$$

describes the production stage at nucleation, and

$$\text{let} \quad H0(X) = H0(X) + 2*GRC*DECG \qquad (4.5.2)$$

the production during subsequent grain growth. Once produced, [H] diffuses under its own pattern of concentration gradients, and its concentration is re-calculated, via $H_1(X)$, at every time T.

52   Periodic Precipitation

In general, [H] will be associated with a diffusion constant of its own, namely D_H. The distribution of [H] in the system is thus controlled not by any reservoir, but by the principal places of its production, and by its subsequent diffusion.

Wherever [H] may be found, it could have an effect on K_S, K_{SP}, and M_{PG0} as well as, in principle, on G_{RC} and R_{SC}. We do not have nearly enough information at hand to model these interactions quantitatively with any degree of confidence. However, a qualitative demonstration can be envisaged without difficulty. For this purpose, we will assume that the principal effect is on the solubility product K_S, which is expected to rise with increasing $H(X)$. If nothing else were done, K_S might at some stage exceed K_{SP}, and that would be absurd. Therefore we should allow K_{SP} to change also; the question is how. We have no reliable information on what happens in practice, and in most cases no information at all. Educated guesses will therefore have to do and, if necessary, some uneducated ones. One way of preventing cross-over would be to allow K_{SP} to change in such a way as to maintain the original difference between these two concentration products. The locally operational values would therefore be written (in Basic) as:

```
        let KS = KS0*[1 + PHF*HO(X)/20]                  (4.5.3)
and     let KSP = KS + [KSP0-KS0],                       (4.5.4)
```

where K_{SP0} and K_{S0} are the originally set values. P_{HF} (suggesting "pH-factor") is a user-determined constant which, by decree, is allowed to vary from 0 to 1. There is nothing subtle about this formulation, and the implied linearity is not intended to signify more than the first term of a Taylor expansion. The factor 20 is there, for no better purpose than to keep the term small, even when P_{HF} has its maximum value. However, events are sensitive to the value of K_S, and things can change substantially as a result.

Another possible way of preventing a K_{SP}-K_S cross-over would be to substitute something like:

```
        let KSP = KSP0*(1 + PHF*HO(X)/20)
```

for equation 4.5.4. See also Section 5.6. Of course, other provisions could be made, and there may be occasions when other, and possibly more complex, interaction models should be explored.

4.6 Reservoir Depletion, Contamination, and Closure.

The diffusion system here envisaged (all along) is represented in Fig. 4.6.1. This shows the diffusion column of length L, (divided into N cells, so as to make $L=N$), as well as reservoirs for the [A] and [B] reagents. When computations are performed in which $A(0)$, $A(L)$, $B(0)$, and $B(L)$ are constants, we are implicitly assuming that the reservoirs are infinite. Indeed, the same assumption is often mistakenly made in the laboratory, where infinite reservoirs are manifestly rare. There was a time when this practice was regarded as a harmless simplification, but it actually has some profound consequences: nourished by infinite reservoirs, no reaction can ever end, and stability is never in sight!

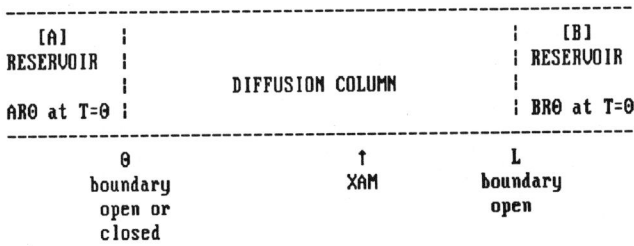

Fig. 4.6.1. Diffusion column, with two reagent reservoirs.

It is therefore well worth while to make provision for reservoir depletion. For this depletion there are actually two reasons. One is the obvious enough; solute, say [A] diffuses out of the [A] reservoir and into the diffusion medium. The other arises from the gradual contamination of the [A] reservoir by in-diffusing [B]. The initial situation is (in Basic):

```
let    A(0)=AR0
let    A(L)=0
let    B(0)=0                                            (4.6.1)
let    B(L)=BR0
```

but in the course of time, $A(0)$ and $B(L)$ are bound to diminish. How fast they diminish depends on (a) the speed of diffusion, and (b) the size of the reservoirs.

It would not have to be so but, for the sake of simplicity, we shall here assume that the [A] and [B] reservoirs are of equal size. Rather than deal with actual dimensions, we once again define a user-determined coefficient, the "reservoir depletion coefficient", R_{DC}. An infinite reservoir would correspond to $R_{DC}=0$. After each pass (from $X=1$ to $L-1$), we then re-set the reservoir concentrations. In Basic:

```
let A0(0)=A0(0)-RDC*DA*[A0(0)-A0(1)]-RDC*DB*B0(1)

let B0(L)=B0(L)-RDC*DB*[B0(L)-B0(L-1)]-RDC*DA*A0(L-1)
```
(4.6.2)

This formulation takes note of the fact that [A] leaves its reservoir at a rate proportional to the concentration gradient at $X=0$. $A_0(0)$ is further diminished by the influx of [B], which is proportional to $B_0(1)-B_0(0)$, the last term being zero.

In entirely similar ways, we can make allowance for the contamination of the two reservoirs by the waste product [H]. Moreover, the amounts of [A] and [B] seeping into opposite reagents will lead to the production of a certain amount of [H] there. We thus have (in Basic):

```
let   H0(0)=H0(0)+RDC*DH*[H0(1)-H0(0)]+RDC*DB*B0(1)

let   H0(L)=H0(L)+RDC*DH*[H0(L-1)-H0(L)]+RDC*DA*A0(L-1)
```
(4.6.3)

If the two reservoirs were unequal, we would simply introduce two depletion coefficients, R_{DCA} and R_{DCB}.

For many kinds of hypothetical experiments, it is desirable and convenient to pre-charge the diffusion medium with one of the reagents, (say) [A], and allow [B] to diffuse in from a reservoir. The question is then whether [A] in the medium should be allowed to be used up by the incoming [B], or whether it should be replenished by influx from an [A] reservoir. In other words, should the boundary at $X=0$ be open (as so far assumed) or closed.

It is a simple matter to make provision for closure. When the boundary at $X=0$ is closed, no concentration gradient can appear at that place, because there is no transport. The condition for that is simply:

```
        let A0(1) = A0(2)
        let B0(1) = B0(2)
        let H0(1) = H0(2).
```
(4.6.4)

Whenever closure at $X=0$ is called for, these lines of code can be inserted into the re-calculation subroutine, to be transacted after each pass.

Section 5.3 will show how reservoir depletion affects the pattern of deposits. Meanwhile, Fig. 4.6.2 gives typical $B_0(0)$ versus time relationships for two depletion coefficients (operative in the specified system). It can serve as a coarse guide for selecting R_{DC}-values for hypothetical experimentation.

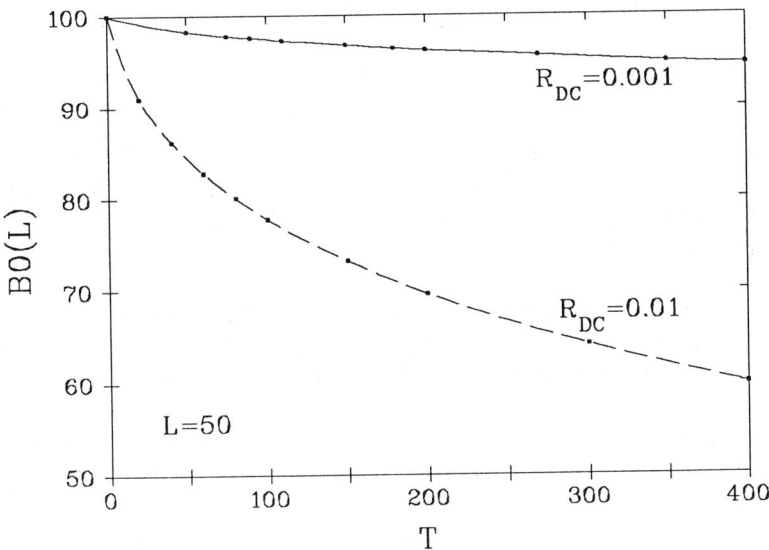

Fig. 4.6.2. Reservoir depletion. Examples for two values of the reservoir depletion coefficient (R_{DC}). Diffusion medium pre-charged with reagent [A] to a concentration 10. $D_A = D_B = 1/6$, $P_{HF} = 0$.

5. HYPOTHETICAL EXPERIMENTATION WITH BINARY SYSTEMS

A microcomputer program which includes the elements discussed above is capable of yielding new insights into the workings of diffusion and precipitation systems. Such software is described in Appendix A. Here it is sufficient to point out that arrangements can be made for the user-control of many operational parameters, so many, indeed, as to make the logical outcome of the multiple choices unpredictable. More precisely, it is sometimes possible to foresee results qualitatively in simple cases, but not in complex ones, and never quantitatively (within the author's experience). In that sense the results represent new research insights, available for comparison with actual laboratory experiments.

There is no way of predicting the reader's special needs and interests; the "experiments" that follow should be understood as typical examples of what can be done. For convenience (only) they are arranged in homogeneous groups on the basis of principal themes, varying only one parameter at a time in order to assess its specific impact on events. Of course, it would be entirely possible to go beyond these constraints, and that is the territory in which results of greatest novelty are likely to be found. Indeed, to encourage and facilitate such experimentation is a principal purpose of this book.

Computer experimentation is a territory in which great freedoms reign. We can, for instance, elect to vary the conditions of an experiment by (say) doubling the value of K_{SP}, while keeping K_S constant, safe in the knowledge that we can do so without ever bothering to think which substance might in practice represent that modified case. Indeed, there may be no such substance. That is fair enough, because the experimentation here conducted is not with substances, but with system structures, parameter patterns, their interactions and their logical consequences. No chemical stockroom resources are thereby taxed, no equipment budgets stretched, no graduate student tempers ruffled. The practical world is not ignored, but some of its limitations are transcended.

58 Periodic Precipitation

5.1 Typical Result Configurations.

The software provisions described in Chapters 1-4 can be linked to "user-interfaces" in many different ways, and there is no compelling logic about the choice of those. Taste and convenience govern the final arrangements, as far as the display and recording of results are concerned. By way of example, the implementation used for the present hypothetical experiments makes provision for three graphic display screens, containing different kinds of status reports about the ongoing processes. In addition, there are conventional arrangements for hard-copy printout (which need not concern us here). The three screens are shown in Fig. 5.1.1. Each consists of two graphs and displays, besides, the most important items of numerical information relating to the experiment in hand. One of these is the current time.

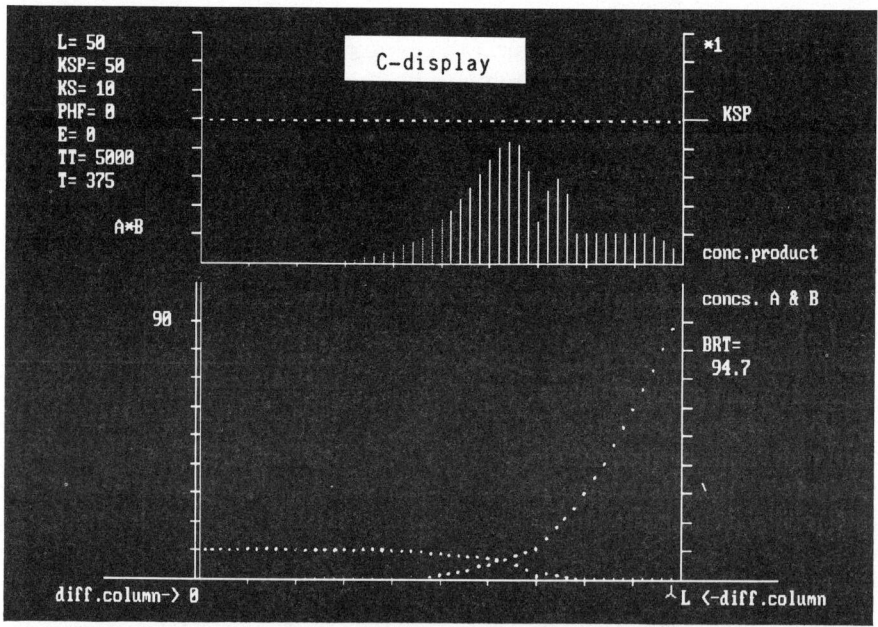

Fig. 5.1.1. Typical result configurations; status displays for monitoring the progress of computation. Examples shown for the parameters listed, together with $D_A=D_B=D_H=1/6$, $R_{DC}=0.001$, $P_{HF}=0$. (a) C-Display: concentrations and concentration products, (b) M-Display: precipitated masses and concentrations of the secondary reaction product, (c) G-Display: grain size and grain count.

Experimentation with Binary Systems

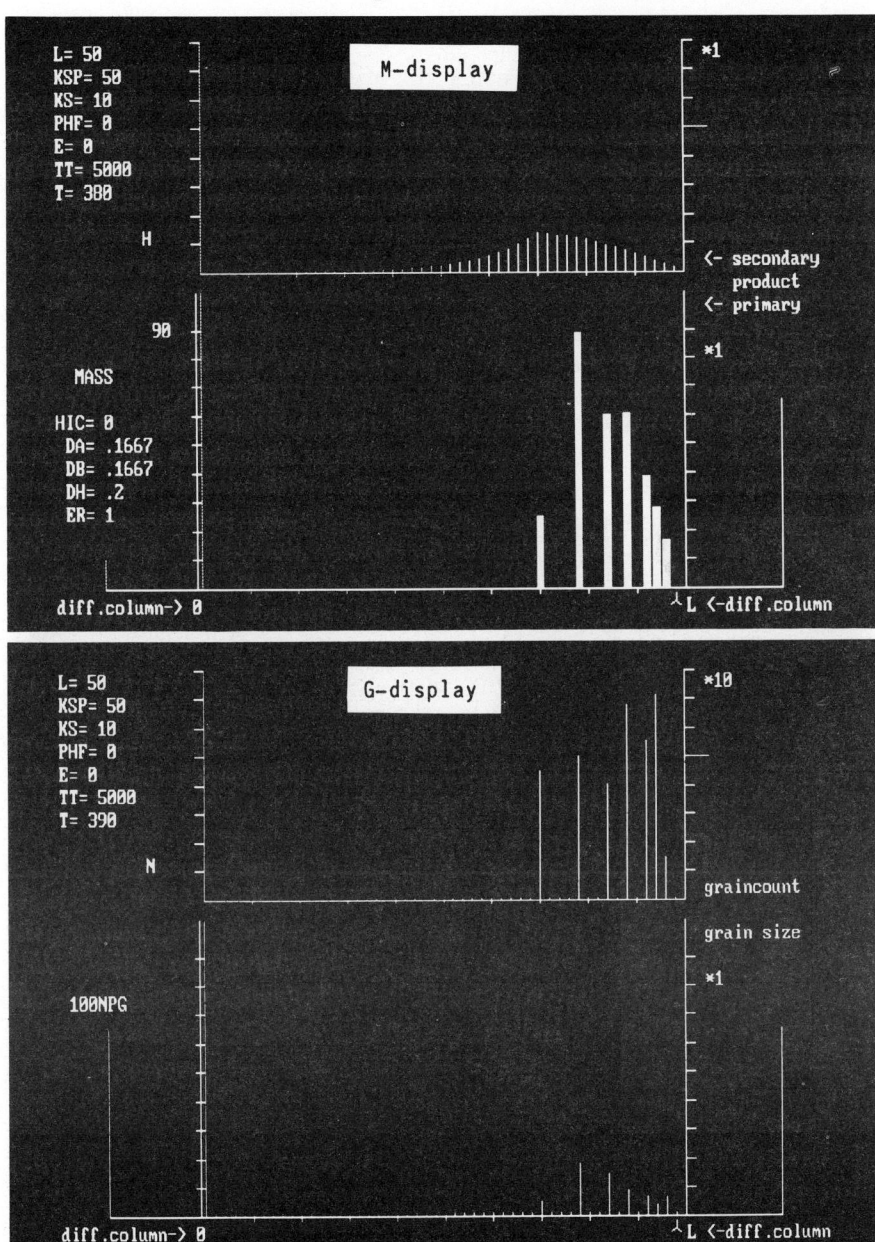

Common features: The double vertical line on the left indicates that, for the particular experiment shown, the boundary at $X=0$ is closed. A single line would have shown it to be open; compare Figs. 5.2.1 and 5.2.2. The reservoir concentration, here only of [B], at the time in question is printed on the right. The position of the inverted Y on the abscissa (here at $X=49$) indicates the extent to which the diffusion column was pre-charged with [A] at the outset. Unless otherwise stated, the information here provided refers to operations in the deterministic mode.

The C-Display (Fig. 5.1.1a): The lower graphs show $A(X)$ and $B(X)$ for the time (here $T=375$) at which action was stopped in order to record this screen image (using PIZAZZ PLUS$^{(TM)}$, a superb product of Application Techniques Inc.) The diffusion column was precharged with [A] to a concentration 10, and this is clearly shown. By the time $T=375$, the concentration of the [B] reservoir has dropped from its original value of 100 to 94.7; compare Fig. 4.6.2.

The upper display concerns itself with the concentration product. On the right, where precipitates and solution co-exist, that product equals K_S over an extended region, as it should. Of course, such a regionally horizontal profile will be seen only when $P_{HF}=0$, i.e. when K_S is really constant. When $P_{HF}>0$ the A.B profile in the quasi-equilibrium region mirrors the X-dependence of K_S, brought about by its H-dependence; see Fig. 5.6.1. In the present case, for $X>46$, we actually have $A(X)*B(X)<K_S$, and precipitates located there must be in the process of re-solution. There is experimental confirmation for this finding (Braterman 1989). At $X=35$, the site of the most recent nucleation, the concentration product has been decremented. In due course the maximum at 37 will disappear, as the product there likewise approaches K_S. The maximum at 32 will be seen to grow until it exceeds K_{SP}, and a new nucleation episode is initiated. Watching the C-Display thus permits the observer to do some informal forecasting as to where the next deposit is likely to be.

The M-Display (Fig. 5.1.1b): The lower part shows accumulated deposit masses, the upper part concentrations of the waste product [H]. It will be seen that $H(X)$ tends to zero, as X tends to L, because the waste product diffuses out into the [B] reservoir, which is here large enough not to be appreciably contaminated thereby. At the X-value which marks the most recent nucleation, namely $X=35$, the concentration $H(X)$ shows a peak, as one would expect. It will be seen that three deposits at the higher values of X are "unresolved"; they should be considered as a single, thick deposit.

The G-display (Fig. 5.1.1c): The lower part gives grain sizes $MPG(X)$, and the upper part $N(X)$. As time goes on, $M_{PG}(X)$ can be seen to change; some grains grow and others re-dissolve. In contrast, the graincount remains unchanged, once a nucleation site is established, because no provision is made for secondary nucleation. However, if at any time $M_{PG}(X)$ comes to be smaller than M_{PG0} (the size of the critical nucleus), then $N(X)$ is set to zero, and $M_{PG}(X)$ to M_{PG0} at that point. In other words, the deposit is then deemed to have disappeared. Secondary nucleation would be possible at that site, as nucleation is at any other place for which $N(X)=0$. Note the increasing grain size towards smaller values of X, not inevitable, but fairly typical; compare Fig. 1.1.1. This tendency becomes more pronounced as time goes on.

The information reported below takes the form of such screens, or else of summarized extracts from them. Unless otherwise stated, the electric field is assumed to be zero.

5.2 Experiments with System Structures.

An aspect often neglected in practical experimentation (e.g. conducted with silica gel as the diffusion medium) is the significance which attaches to the length (L) of the diffusion column. That length may determine how quickly one of the reagents reaches exhaustion. It is a simple matter to conduct a computer-experiment which provides a basis of comparison.

Figs. 5.2.1 and 5.2.2 relate to an experiment in which the diffusion column is pre-charged with [A] to a concentration $A_{M0}=10$. The column can be closed at $X=0$, or else open, i.e. in communication with the [A]-reservoir. When it is open, we make the reservoir concentration $A_{R0}=10$, thereby simulating (not precisely, of course, but sufficiently well for present purposes) an infinite reservoir. During the initial stages of the process, there is no difference, and none is expected, but after longer times, the differences are pronounced. Fig. 5.2.1 shows them in terms of concentration products. In Fig. 5.2.1a, the boundary is closed, and [A] is tending to exhaustion. The concentration product is actually falling, and can never again reach K_{SP}. Further consumption of the remaining [A] is due to deposit growth at ever diminishing rates. Note the extended equilibrium region. In Fig. 5.2.1b, there is access to the reservoir, and this guarantees a continuing supply of [A]. This permits the build-up of a new product maximum at $X=10$. As it happens (with the parameters chosen) that maximum does not grow to K_{SP} either, but it comes very close to it. A very slight change of parameters (e.g. a slightly smaller value of R_{DC}) would lead to an additional precipitate there.

62   Periodic Precipitation

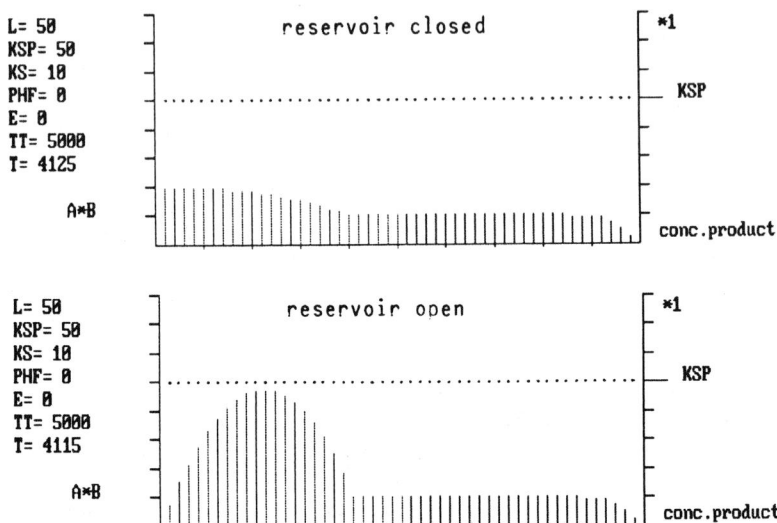

Fig. 5.2.1 Effect of reservoir closure on the concentration product profile. Example for the parameters listed, together with $D_A=D_B=1/6$, $R_{DC}=0.001$, $P_{HF}=0$, $A_{R0}=10$, $A_{M0}=10$.

Fig. 5.2.2 shows the differences in terms of deposited masses. In Fig. 5.2.2b, the deposit at $X=20$ is just about twice as massive as that for the case in which the boundary is closed. This is, of course, a consequence of the [A]-supply that is here derived from the [A]-reservoir; such a supply could alternatively be derived from a longer diffusion column, specifically from regions in which precipitation has not yet taken place.

5.3 Experiments with Reservoir Concentration Parameters.

Under this heading we shall conduct two simple tests, both designed to illustrate the importance of reservoir design and specifications. In the first, we shall monitor how the precipitation pattern depends on the reservoir depletion coefficient (R_{DC}), other things being equal. Fig. 5.3.1 illustrates this point for a partricular case, taking R_{DC} in turn as 0, 0.006 and 0.06. We would not expect any differences for small values of T, but after longer times there are obvious differences in the density of deposits.

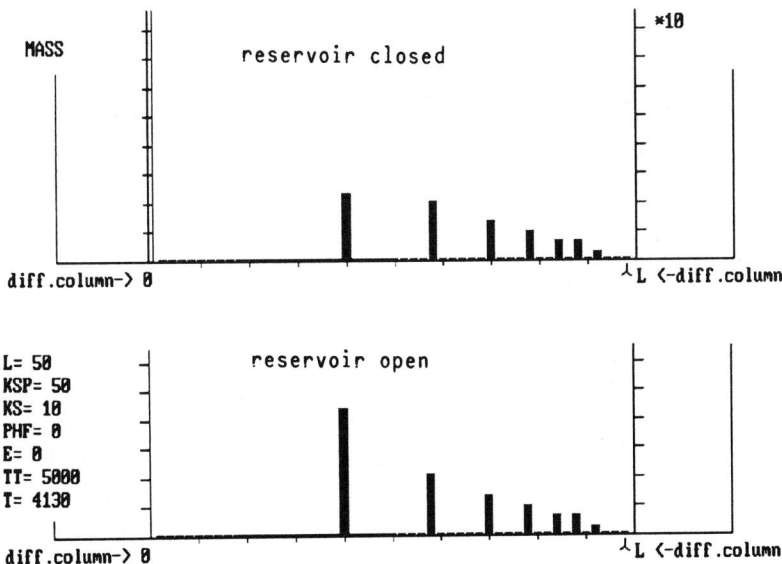

Fig. 5.2.2. Effect of reservoir closure on the deposit mass profile. Example for the parameters listed, together with $D_A=D_B=1/6$, $R_{DC}=0.001$, $P_{HF}=0$.

and their position. Differences in the growth rates (not shown) can also be recorded. This illustrates the folly of an (alas, widespread) malpractice in laboratory experiments, namely that of describing the conditions of the diffusion medium with elaborate care, without paying proper attention to the reservoir parameters. In particular, it makes no sense to discuss "spacing laws", without tying the findings to specific parameter choices.

Systems in which deposits are formed "late" in the proceedings are, of course, particularly sensitive to the choice of R_{DC}. Thus, in a variant of the above experiment, the medium was pre-charged with [A] only $X_{AM}=25$ which, of course, postpones the first precipitation. For $R_{DC}=0$, four deposits are recorded at $T=1250$, located at $X=21$, 30, 37, and 38; for $R_{DC}=0.06$, only two, located at $X=25$ and 36.

In the second experiment, we make a back-reference to Section 2.4, where it was shown that the first deposit is expected to be formed in the middle of a (previously empty) diffusion column, even though the reservoir concentrations might be grossly unequal. The question is whether deposits subsequently formed are also symmetrical in such circumstances. Fig. 5.3.2a shows that the first deposition event is symmetrical, but subsequent ones are not. Recall once again that discussions about the location of the first concentration product maximum need to concern themselves only with diffusion, whereas those about subsequent deposits involve, besides diffusion, also grain growth and grain re-solution. The situation, in fact, ceases to be symmetrical immediately after the first event. In the specific example shown, the second deposit forms at $X=33$, without any corresponding formation at 17. A third and much later deposit (not shown) grows at $X=41$, far on the [B] side, where we might not expect it. This formation is nevertheless reasonable. The small amount of [B] that is able to penetrate to low values of X, past two already established and still growing deposits, is unable to promote a new precipitation; the much larger amount of [A] that is able to penetrate to high values of X certainly is. Fig. 5.3.2a gives a status report at $T=2000$, and symmetry is not restored by later events.

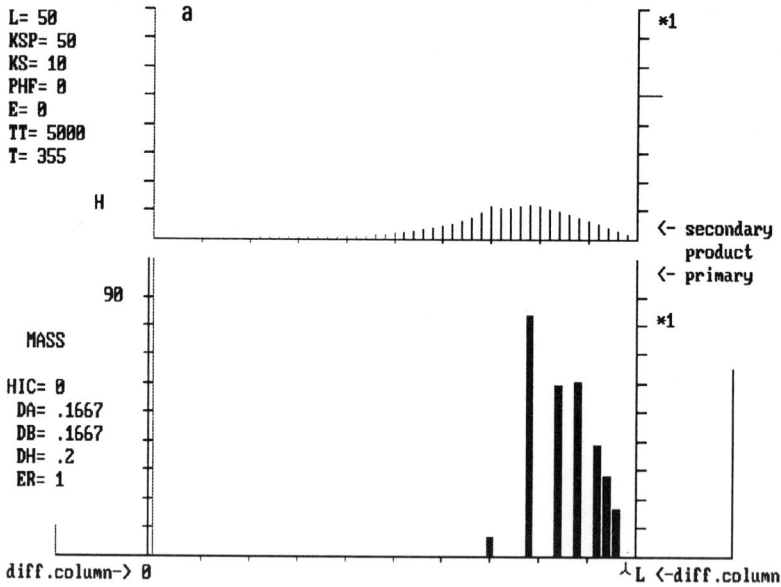

Fig. 5.3.1. Effect of the reservoir depletion coefficient on the deposit mass profile. Example for $L=50$, $K_{SP}=50$, $K_S=10$, $P_{HF}=0$, $E=0$. (a) $R_{DC}=0$, $T=575$, (b) $R_{DC}=0.006$, $T=575$, (c) $R_{DC}=0.06$, $T=580$.

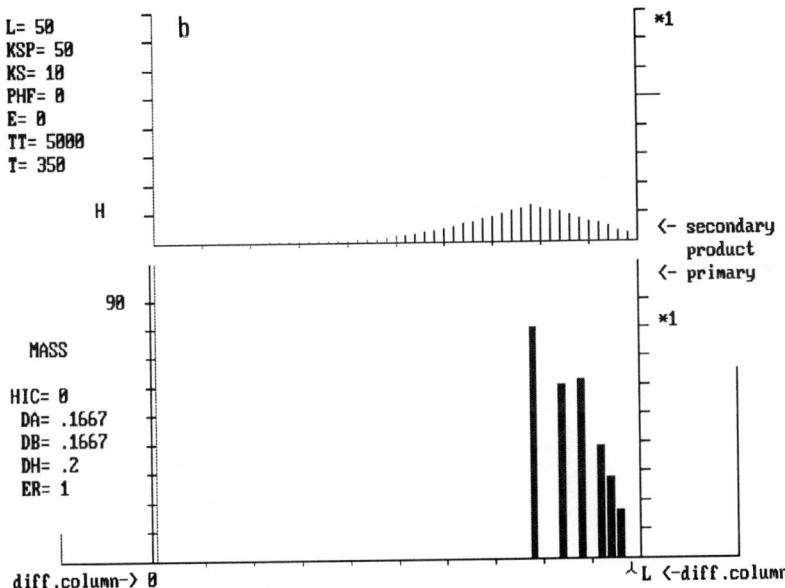

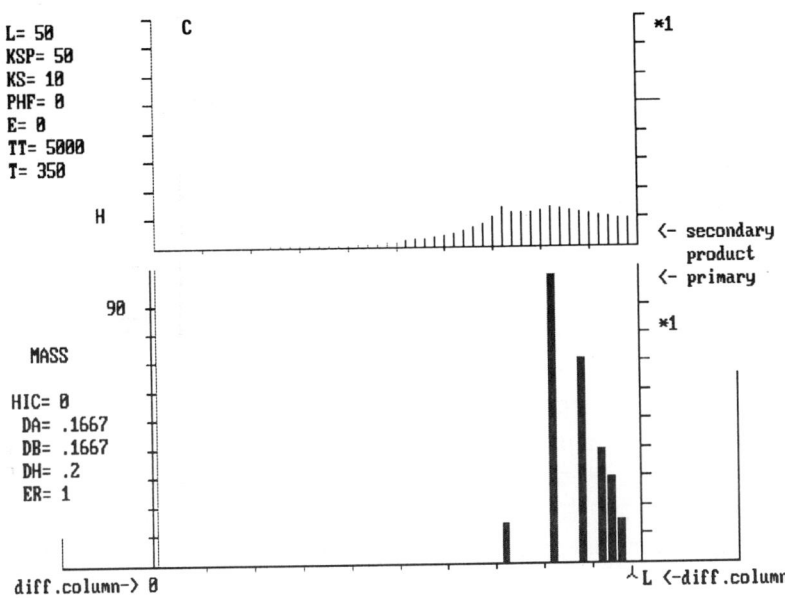

As one might guess, the symmetrical position of the first deposit in Fig. 5.3.2a is entirely dependent on the suspension of the stoichiometry considerations, achieved by setting $E_R=1$; see also Section 5.7. When stoichiometry requirements are enforced, as they are in Fig. 5.3.2b ($E_R=0.1$), even the first deposit is off-center.

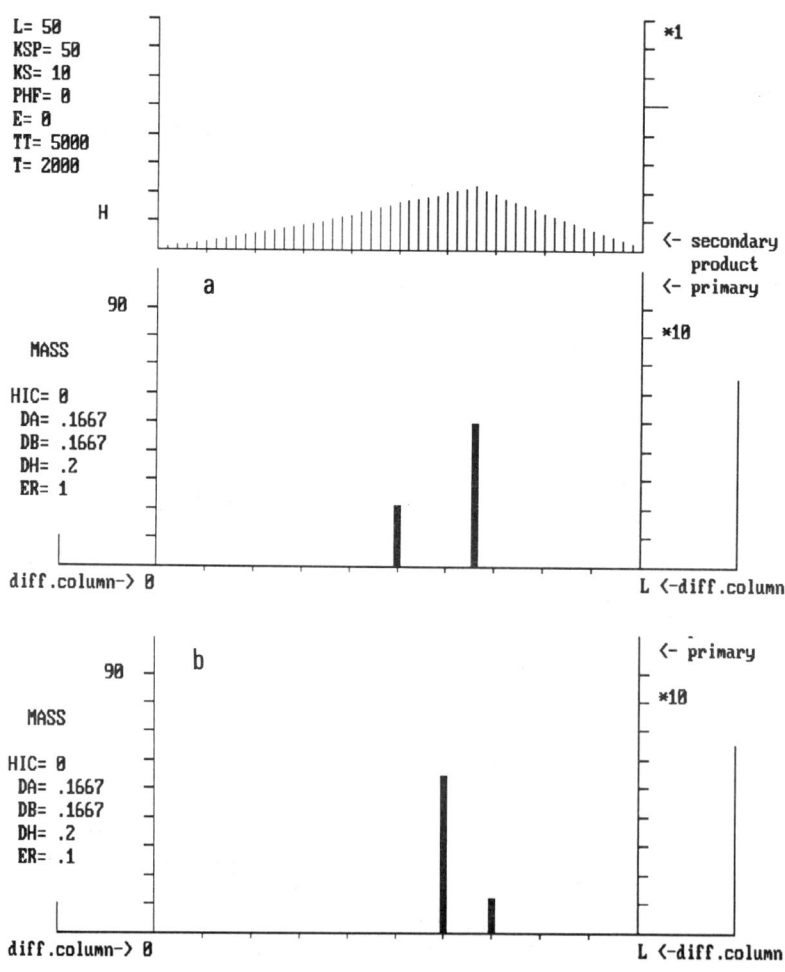

Fig. 5.3.2. Effect of unequal reservoir concentrations on the deposit mass profile. Example for $L=50$, $A_{R0}=100$, $B_{R0}=30$, $K_{SP}=50$, $K_S=10$, $P_{HF}=0$, $T=2000$. (a) $E_R=1$, (b) $E_R=0.1$.

5.4 Experiments with Diffusion Coefficients.

Two straightforward test opportunities present themselves. We could, for instance, ask: To what extent can higher diffusion constants make up for lower terminal concentrations? By now, we actually know the answer: no simple compensation is possible, just because growth as well as diffusion is involved. As a result, the diffusion constants affect the deposit configuration in complex ways that are rarely predictable on a priori grounds.

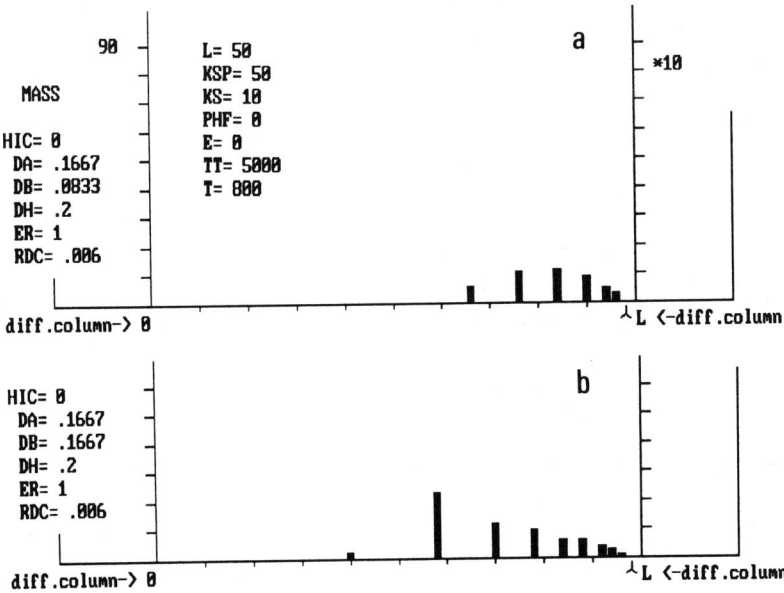

Fig. 5.4.1. Effect of diffusion constants on the deposit mass profile. Example for $L=50$, $K_{SP}=50$, $K_S=10$, $P_{HF}=0$, $E=0$,; access to the [A] reservoir open. (a) $D_A>D_B$, (b) $D_A=D_B$, (c) $D_A<D_B$.

Fig. 5.4.1 gives an example, a case in which one form of asymmetry (pre-charging of the medium with [A]) is built-in, while another arises (in a and c) from the unequal values of D_A and D_B. Fig. 5.4.1b, for which $D_A=D_B$, may be regarded as the standard of comparison. It is ordinarily assumed that deposits appear in the order of their positional sequence, but this is not always the case. Thus, early deposits appear first at high X-values and then at lower ones, but the peak at $X=25$ appeared after that at $X=20$. Many observations of this kind can be found in the literature, events in which late-forming deposits "insert themselves" somewhere in an already established patter of deposits, and disturb simplistic notions of its regularity. The results shown in Fig. 5.4.1 make the importance of D_A and D_B perfectly clear. Once again, there are times when simple and qualitative predictions based on intuitive understanding come true; complex and quantitative ones almost never do.

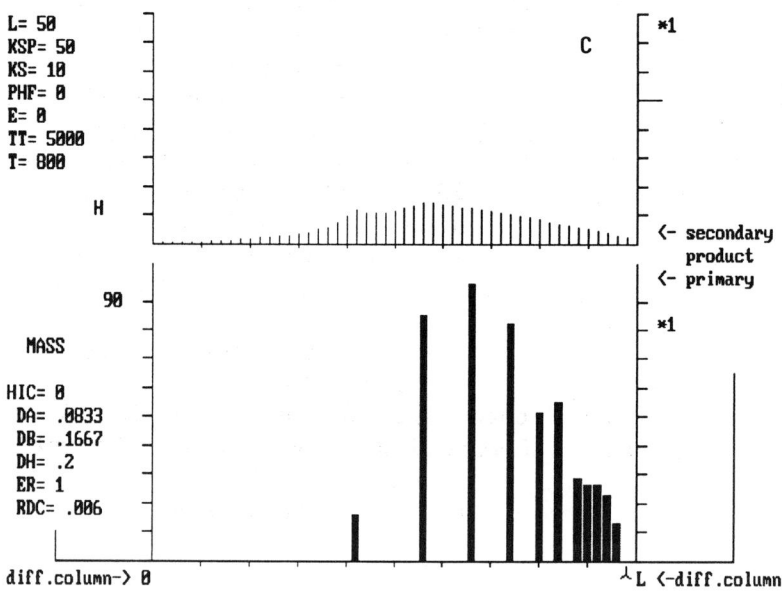

Fig. 5.4.1c Effect of diffusion constants on the deposit mass profile. $D_A < D_B$. Note change of scale.

5.5 Experiments with Solubilities and Precipitation Thresholds.

It is simple enough to foresee that higher values of K_{SP} (other things being equal) will lead to fewer deposits (wider deposit spacings), and that higher values of K_S (again, other things being equal) will lead to more closely spaced precipitates. It is quite another matter to predict where deposits will appear, when and of what mass. For that one would have to enlist the services of a talented clairvoyant, were it not for the fact that the computer is more reliable (not to mention cheaper).

Table 5.5.1, a summary taken from the corresponding M-Displays, gives a series of results, derived from experiments in which K_{SP} was kept constant, and K_S gradually increased. The inter-deposit spacing is indeed found to diminish. For a specific example [$L=50$, $X_{AM}=49$, $A_{M0}=10$, $B_{R0}=100$, $E=0$, $D_A=D_B=1/6$, $P_{HF}=0$, $R_{DC}=0.001$, $M_{PG0}=0.01$, $E_R=1$] the numbers confirm the prediction.

Table 5.5.1
Site Occupancy as a Function of K_S for $K_{SP}=50$.
Percentage of occupied sites at $20=<X<=50$. Time=150.

K_S	% occupancy
2	27
20	37
40	43
45	63

Table 5.5.2
Site Occupancy as a Function of K_{SP} for $K_S=10$.
Percentage of occupied sites at $40=<X<=50$. Time=150.

K_{SP}	% occupancy
15	90
20	80
50	60
100	40
200	20
300	0

At the same time, as K_S approaches K_{SP}, the deposit masses diminish sharply. This is due to the diminishing number of grains formed, in accordance with equation 3.3.2 in which, of course, D_{EC} is governed by the extent to which K_{SP} exceeds K_S. The grain size turns out to be larger for higher values of K_S, but not large enough to compensate for the smaller $N(X)$ values.

Corresponding experiments can, of course, be performed by keeping K_S constant and varying K_{SP}. As K_{SP} increases, ever larger super-

70   Periodic Precipitation

saturations are required for precipitation, and deposits become increasingly rare, as shown in Table 5.2.2. Oddly positioned "late arrival" deposits were again observed in the course of these runs, though at various times considerably longer than $T=150$.

5.6 Experiments with a Secondary Reaction Product.

The number of configurational possibilities (in combination with all the other parameters discussed) is, of course, enormous. Without aiming to prove any particular point, we shall simply illustrate typical patterns of interaction between the primary product [AB] and a secondary product [H] that tends to make [AB] more soluble, to an extent controlled by the software user.

Since [H] is produced in proportion (indeed, 1:1) to the mass of deposited [AB], we will expect to find its effects greatest in the neighborhood of the greatest deposit concentrations. On the other hand, the effects are not limited to those regions, because [H] diffuses under its own pattern of concentration gradients, with a different diffusion constant. Because it is tempting to regard H as, for instance, the concentration of hydrogen ions, the diffusion constant D_H (used below) has been given a value somewhat higher than D_A and D_B. Of course, other choices could be made, e.g. assigning a small value to D_H when [H] is a heavy, large molecule. In order to remain within the realm of "small effects", the range of P_{HF}, which governs the interaction with K_S (see Section 4.5) will be restricted to $0=<P_{HF}<=1$

In Fig. 5.5.1a which applies to $P_{HF}=0$, we noted the extended quasi-equilibrium region over which deposits and solution co-exist. The observation of some such region is typical when K_S is constant, i.e. when it is not influenced by $H(X)$. When it is so influenced, e.g. in accordance with equations 4.5.3 and 4.5.4, we will expect to see a contour which roughly mirrors the local values of K_S. Fig. 5.6.1 (for $P_{HF}=1$) shows this to be so.

The effect of [H] on grain size and grain count can be substantial. In terms of detail, much depends on the speed with which [H] reaches new potential nucleation sites, which means also that much depends on how rapidly in succession new deposits are being formed. The more rapid that succession, the smaller is the effect of [H] on nucleation, and the less important the value of P_{HF}.

Because equation 4.5.4 leaves K_{SP}-K_S constant, we would not immediately expect any major effect of [H] on the grain count, but because all the parameters interact with one another in a complex way, there is such an effect. The left-hand side and central columns of Table 5.6.1 (in which only X-values are listed for which deposits are found) shows this for a particular case, described by the parameter choices: $L=50$, $K_S=10$, $K_{SP}=20$, $A_{R0}=100$, $B_{R0}=100$, $R_{DC}=0.001$, $D_A=D_B=1/6$, $D_H=0.2$, $M_{PG0}=0.01$, $E_R=1$, $T=2750$, closed [A] reservoir.

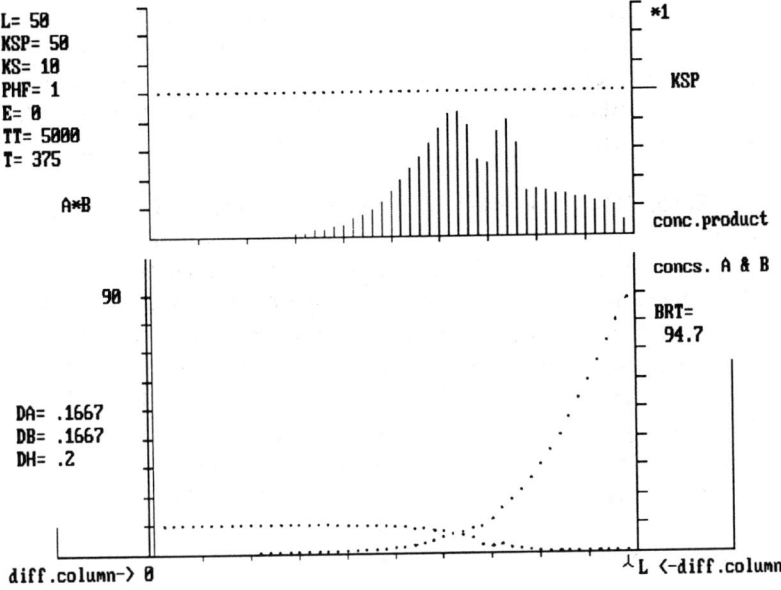

Fig. 5.6.1. Interaction with a secondary product. A concentration product profile for $P_{HF}=1$. Example for $D_A=D_B=0.1667$, $D_H=0.2$, $E_R=1$; equations 4.5.3 and 4.5.4.

Note that $M_{PG}(X)$ is routinely assigned the value M_{PG0}, corresponding to the size of the critical nucleus, whenever $N(X)=0$; no physical meaning intended.

All the deposits at $X>=35$ exist before $T=400$, and the mass distributions are very similar. However, even in these early formations, $N(X)$ is always smaller when [H] interacts than when it does not. All the deposits for smaller values of X represent later formations, and the differences between $P_{HF}=0$ and $P_{HF}=1$ runs are then much greater, as one would expect. At $T=2750$, the time at which these results were recorded, the system has reached quasi-stability. Surprisingly, considering that an increasing K_S represents an increasing solubility, the total masses deposited under the two conditions are by then about equal.

Table 5.6.1
Effect of Secondary Reaction Product..

X	$P_{HF}=0$			$P_{HF}=1$, equ. 4.5.4			$P_{HF}=1$, equ. 4.5.5		
	$N(X)$	M_{PG}	Mass	$N(X)$	M_{PG}	Mass	$N(X)$	M_{PG}	Mass
16				451	0.29	129			
18							459	0.31	154
20	505	0.39	198						
27				500	0.44	221			
29	505	0.39	198						
30							508	0.48	246
34				544	0.25	138			
35	449	0.29	131	422	0.08	32			
37							629	0.24	149
38							480	0.07	32
39	499	0.20	100	466	0.21	99			
42	405	0.16	63	380	0.17	64	613	0.17	104
44	675	0.09	63	618	0.01	63			
45							736	0.10	72
46	584	0.64	38	505	0.08	39			
47				621	0.01	9	709	0.32	23

Note that only the locations at which deposits actually occur are listed. $N=50$.

It goes without saying that the above results indicate only trends which are associated with the particular pattern of assumptions made. A different way of handling the [H]-dependence of K_{SP}, for instance, would lead to different deposit distributions, each unpredictable by purely intuitive arguments. Thus, the right-hand side of Table 5.6.1 lists (for $P_{HF}=1$) the deposits obtained when equation 4.5.5 is used in place of equ. 4.5.4. The presence of [H] also promotes re-solution, and by the time $T=$ (say) 4500, the deposit at $X=47$ has already disappeared, and the two next ones are in an advanced state of re-solution. The essential non-linearity of the system can be demonstrated by showing that the results for $P_{HF}=0.5$ interpolate very well at small values of T, and not at all at high values.

Fig. 5.6.2 shows (for a typical case) that the concentration product can substantially exceed the originally set value of K_{SP}, when $P_{HF}>0$. Note again the slope of the product contour in the extended co-existence region. (Compare Fig. 5.7.1a which records an apparently similar observation, but one encountered for very different reasons.)

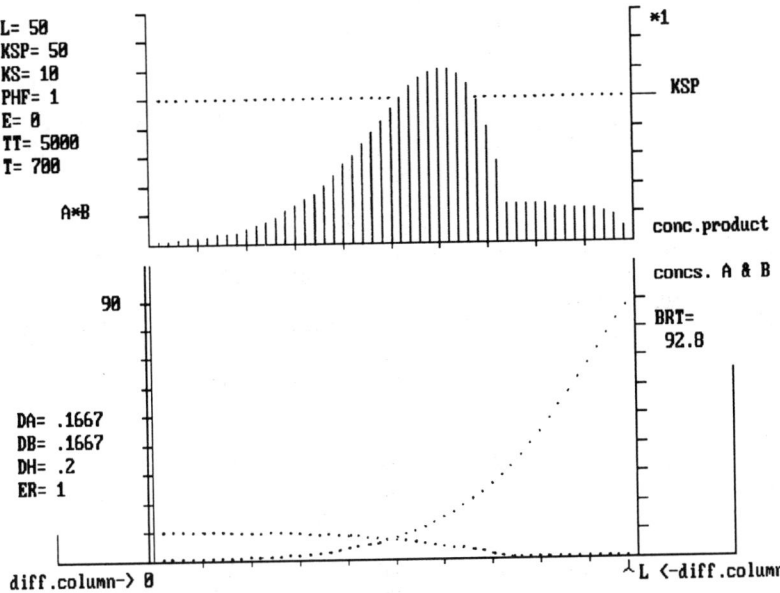

Fig. 5.6.2. Example of substantial interaction between K_{SP} and the secondary product [H]. Note that the concentration product can now exceed the originally set value of K_{SP} (horizontal dotted line). Example for $D_A=D_B=0.1667$, $D_H=0.2$, $E_R=1$; equations 4.5.3 and 4.5.5.

5.7 Experiments with the Stoichiometry Conditions.

In all the experiments discussed so far (unless otherwise stated) the value of the E_R parameter (see Section 3.2) was set to 1, which means that the stoichiometry requirements for nucleation have been ignored. Such a thing is permissible only for simple demonstration purposes, and is even desirable in that context, because it ensures that nucleation occurs precisely when $A(X).B(X)=K_{SP}$, meaning either at the original K_{SP} value (namely K_{SP0} when $P_{HF}=0$), or else at the locally augmented value when $P_{HF}>0$. However, in all realistic cases we must set $E_R<1$. This means that the concentration product $A(X).B(X)$ can grow beyond K_{SP}

(even when $P_{HF}=0$) without precipitation, until the stoichiometry requirements are finally satisfied somewhere. Fig. 5.7.1, for $E_R=0.1$, gives a typical example, showing the concentration product contour just before and just after a precipitation event. In Fig. 5.7.1a, the $A(X).B(X)>=K_{SP}$ requirement is already fulfilled over a wide range of X-values, but no precipitation has occurred, because the concentrations of [B] are far higher than those of [A]. In the system under review, ($A_{M0}=10$, $X_{AM}=49$, $B_{R0}=100$) there is, of course an equality point, but it occurs (at the time shown) where $A(X).B(X)<K_{SP}$. The region over which the concentration product exceeds K_{SP} moves towards the left, and when precipitation finally occurs (Fig. 5.7.1b), it does so necessarily on the left-hand side of that excess bulge. At that point, the local concentrations are depleted immediately, and the remaining points on the right follow later, their solute contents contributing to growth, rather than new precipitations.

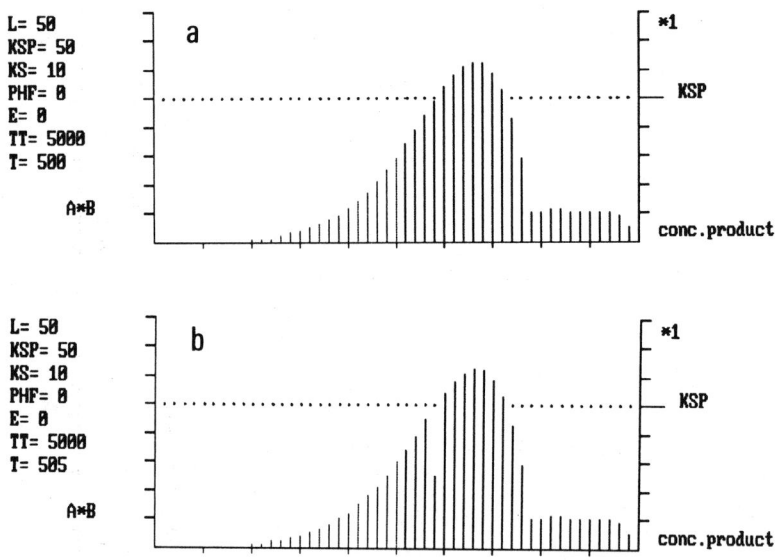

Fig. 5.7.1. Effect of the "equality range" parameter (E_R) on the concentration product profile (a) just before, and (b) just after a precipitation event. $D_A=D_B=0.1667$, $D_H=0.2$, $E_R=0.1$.

It goes almost without saying that by controlling the conditions under which precipitation can occur, the E_R parameter exerts an enormous influence over the entire precipitation pattern. Fig. 5.7.2 shows this. As we qualitatively expect, smaller values of E_R make precipitation more difficult, because they impose increasingly stringent conditions on it.

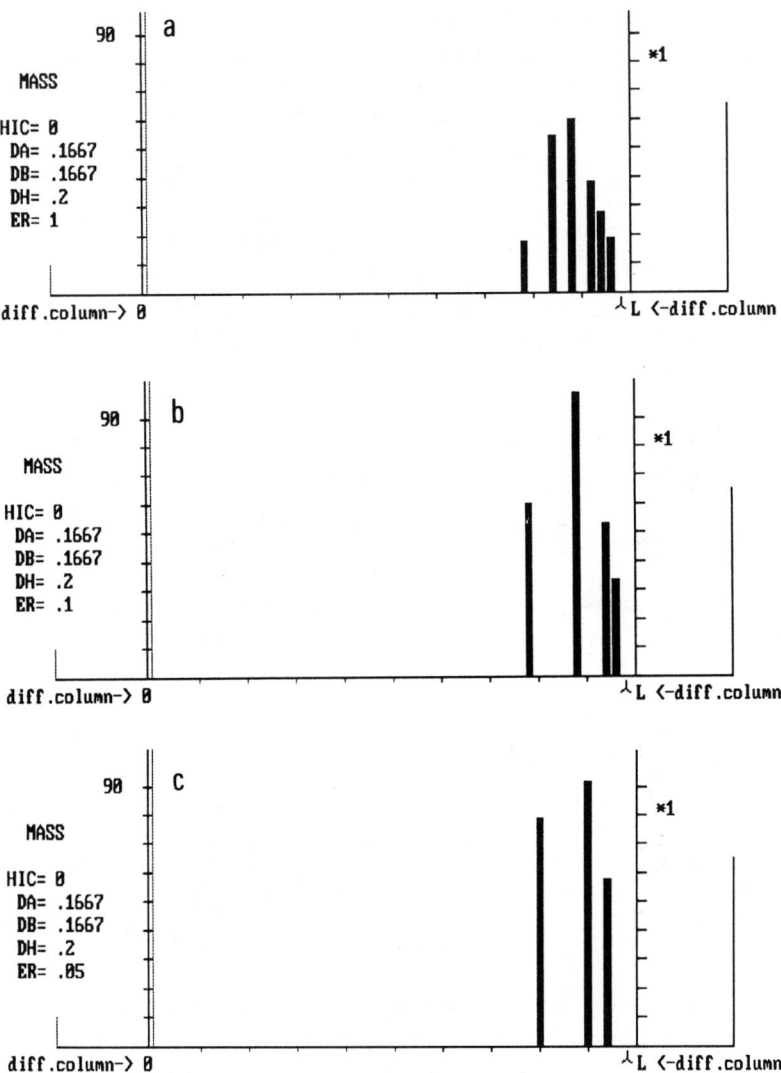

Fig. 5.7.2. Effect of the "equality range" parameter (E_R) on the mass deposition profile. $L=50$, $K_{SP}=50$, $K_S=10$, $P_{HF}=0$, $E=0$, $T=200$ (a) $E_R=1$; effectively no stoichiometry requirements, (b) $E_R=0.1$; (c) $E_R=0.05$.

5.8 Experiments with Electric Fields.

The idea of controlling precipitation events externally by applying electric field is attractive, but we shall have to regard the results of the present experimentation as qualitative, rather than quantitative. This caution is prompted by the approximations made in Section 1.3. The field was there assumed to be "high" and uniform, meaning that all local space charges (as well as local changes of conductivity) were neglected. This approximation applied not only within the bulk of the diffusion medium, but also at its boundaries (electrodes). It remains to be seen how closely laboratory situations which satisfy these conditions can be approached. For experimental records of periodic precipitation in the presence of fields, see Miyamoto (1937), George and Vaidyan (1981 a and b, 1982 a and b), Happel et al. (1929 a and b), Kisch (1929 a and b), Dhar and Chatterji (1925), and Christomanos (1950). However, though there are plenty of "references" to field effects, there is no known record of detailed observations on any well-characterized system, a regret in one sense, but a fine research opportunity in another.

Two more cautions. The simplified field implementation of Section 1.3 does not correctly reflect the effect of electric fields on reservoir depletion. It fails to do so, because it neglects (electrolytic) field currents altogether. It is therefore best, for computer experimentation with fields, to set the reservoir depletion coefficient (R_{DC}) to zero, signifying inexhaustible reservoirs. A second problem is connected with the question whether the [A] reservoir is open to the diffusion column or closed at $X=0$. With a little ingenuity, it would would certainly be possible to introduce electrodes into a closed-off system, and use them for the application of an electric field. However, the more normal situation for field experimentation would be to envisage an open boundary between the [A]-reservoir and the diffusion medium (see Section 4.6).

Fig. 5.8.1 shows field effects for a specific case, at $T=3000$, by which time some deposits at the high end have already re-dissolved. It will be recalled (from Section 2.3) that a positive field drives [A] towards the right, and [B] towards the left. One might naively believe that an electric field should affect only the position of deposits, but this cannot be: positions affect concentration contours and those, in turn, affect the formation of new deposits. Once again, we have here an example of a system described by highly interactive parameters. Of course, higher fields have more drastic results. Thus, for instance, $E=1$, for the same system, fills the diffusion column with a large number of almost equidistant deposits, whereas $E=-1$ allows only a single deposit to form. The implicit control possibilities are obvious enough.

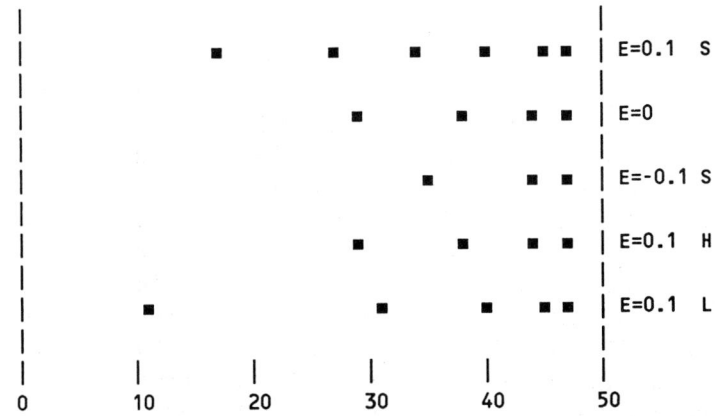

Fig. 5.8.1 Effect of electric fields on the deposit distribution.
Axample: $A_{M0}=10$, $X_{AM}=49$, $D_A=D_B=1/6$, $K_S=10$, $K_{SP}=50$, $R_{DC}=0.006$, $P_{HF}=0$, $E_R=0.1$, $M_{PG0}=0.01$, $T=3000$.
Static field: (S).
Square-wave AC field: half-period $T_{HP}=1$ (H)
　　　　　　　　　　　half-period $T_{HP}=10$ (L).

It is actually quite simple to make the field alternating, instead of static, at any rate alternating in the form of a square wave. All we would have to do is to interpose a few lines of code in front of equation 1.4.10, which is for [A], and the corresponding equation for [B], namely (in Basic):

```
if T/2 = INT(T/2) then    ! meaning if T is even
    let E =  E0
  else
    let E = -E0
end if.
```

where E_0 is the field value originally set. The line designated H (for "high frequency") in Fig. 5.8.1 gives an example, for comparison with the results for $E=0$. Surprisingly, considering the essential non-linearity of things, the alternating field makes no difference (at any rate, in this case), but the failure to observe an effect may be (and, indeed, is) due simply to the fact that the frequency implied by the above code is "very

78   Periodic Precipitation

high" (E changes sign at every T). To lower the frequency, we could define a half-period H_{PI}, as below, and then continue more or less as before. In Basic:

```
if HPI = INT(T/THP) then
    if HPI/2 = INT(HPI/2) then
        let E = E0
    else
        let E = -E0
    end if
end if.
```

If we now make T_{HP} equal to (say) 10 (instead of 2, as above) there certainly is an alternating field effect, and a drastic one at that (see Fig. 5.8.1). One would expect to find the same sort of frequency dependence in a laboratory case. Naturally, the question of where "low" frequencies end and "high" frequencies begin depends on the values adopted for the diffusion constants.

There is also something to be said for applying a static field, but only temporarily, e.g. in order to influence the position of a specific deposit. Of course, this can easily be done. Table 5.8.1 gives examples of results, for a specific deposit in each of two systems.

Table 5.8.1
Effect of temporary electric fields on deposit positions.
System as for Fig. 5.8.1.

$K_S=10$, $F_{AT}=120$, $F_{RT}=140$		$K_S=30$, $F_{AT}=180$, $F_{RT}=200$	
E	X	E	X
0	39	0	37
-0.25	38	-0.5	35
-0.35	37	-1.0	34
-0.50	36		

Here F_{AT} are the field application times, and F_{RT} to field removal times, parameters which, of course, play no role in Fig. 5.8.1. The effects described are all for negative fields; in this particular case, positive fields of the same magnitude (and applied at and for the same relatively short time) have no effect.

Experimentation with Binary Systems   79

The results in Table 5.8.1 represent the consequences of a single field application, but it is clear enough that the repeated application of fields at well-chosen moments (alas, determined empirically) would lead to a substantial measure of control over the precipitation pattern.

From a programing point of view, it is desirable to avoid unnecessary slow-downs. Thus, in a comprehensive software package, the subroutines which call for user input relating to electric fields (e.g. of T_{HP}, F_{AT} and F_{RT}) would be called only when E (which is set first) is not set to zero.

5.9 Experiments in the Probabilistic Mode.

As already noted in Section 4.4, experimentation in the probabilistic mode ($D_{NC}<1$) calls for multiple runs; no specific run is significant.

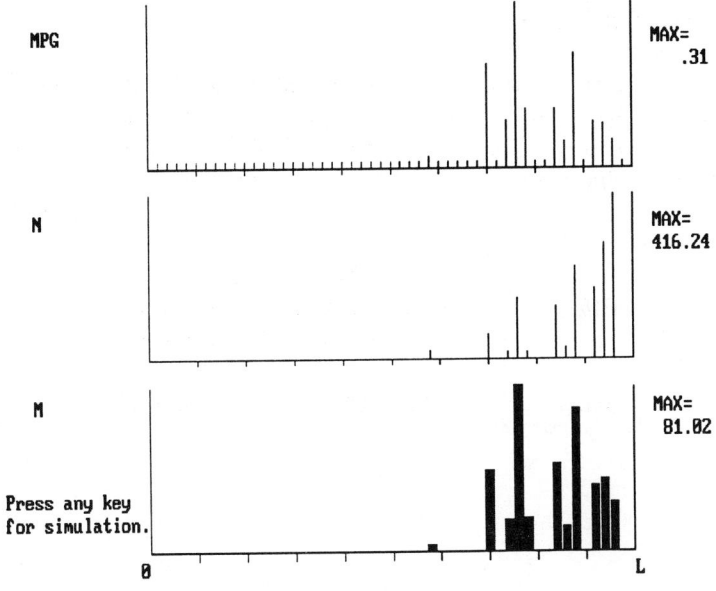

Fig. 5.9.1. Results of an experiment in the probabilistic mode, averaged over 10 runs to time $T=520$. Method A, based on equation 4.4.1. Example described by $D_{NC}=0$, $L=50$, $K_S=10$, $K_{SP}=50$, $E_R=0.1$, $X_{AM}=49$, $D_A=D_B=0.1667$, $M_{PG0}=0.01$, $A_{M0}=10$, $R_{DC}=0.006$, $B_{R0}=100$.

80   Periodic Precipitation

Under such conditions, no purpose would be served in watching a video display of the proceedings. Turning it off not only protects us from boredom, but actually saves time, since the programs run a great deal faster without it.

Fig. 5.9.1 gives a simple demonstration of a multiple run outcome, available for comparison with 5.9.2, which shows the corresponding outcome of a single, deterministic run. It will be seen that the probabilistic deposits do tend to have a real thickness, whereas they tend to have mere locations (rather than real thicknesses) under deterministic conditions.

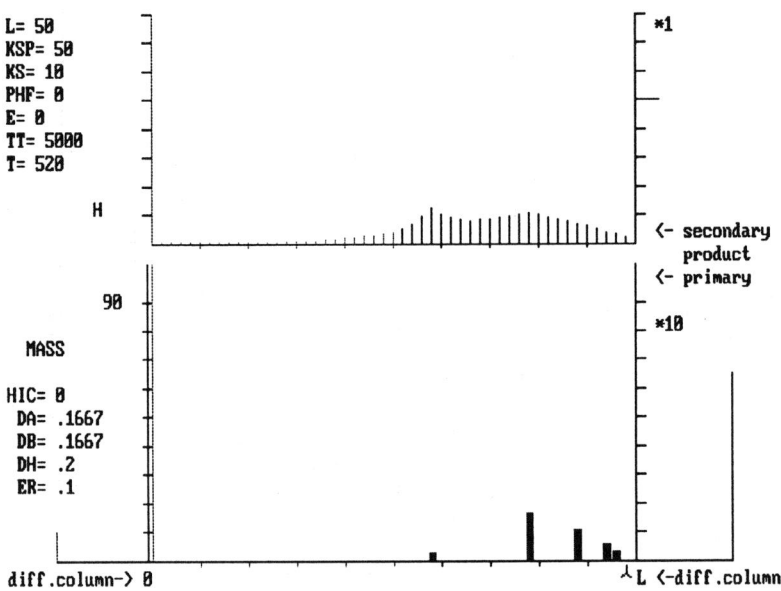

Fig. 5.9.2. Results of an experiment in the deterministic mode; single run to $T=520$. Conditions as for Fig. 5.9.1, except for $D_{NC}=1$.

Section 4.4 actually discussed two methods of introducing the random element into the proceedings, and Fig. 5.9.1 was based on Method A, using $D_{NC}=0$. Fig. 5.9.3 gives an example obtained with Method B, but for $D_{NC}=0.5$, since $D_{NC}=0$, when used with this method, would inhibit nucleation altogether. It is, of course, quite possible to translate the the results, here averaged over 10 runs, into a simulation, one that is shown in Fig. 5.9.4. Such a simulation could be based on the deposit masses, but since we are in any event bound to deal with a dot pattern, it seems more appropriate to base it on $N(X)$, the number of grains nucleated in the vicinity of X. The procedure yields a pleasant visualization though, of course, no new results.

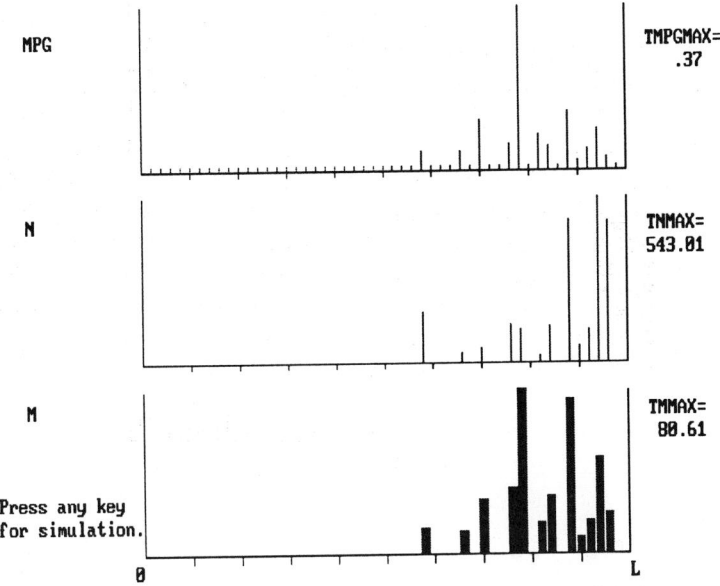

Fig. 5.9.3. Results of an experiment in the probabilistic mode, averaged over 10 runs to $T=520$. Method B, based on equation 4.4.2. Conditions as for Fig. 5.9.1, except for $D_{NC}=0.5$.

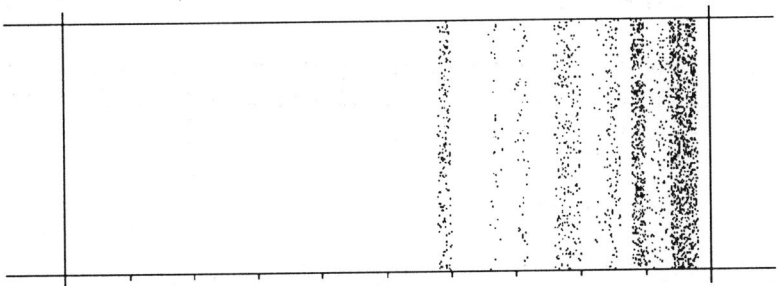

Fig. 5.9.4. Results of an experiment in the probabilistic mode, averaged over 10 runs to $T=520$. Conditions as for Fig. 5.9.3. Simulation based on $N(X)$.

6. HYPOTHETICAL EXPERIMENTATION WITH MONOMER SYSTEMS

6.1 Periodic Precipitation in Monomer Systems.

In all the discussions of Chapters 1-5 the underlying assumption was that two reagents, [A] and [B], come together by diffusion, and form an insoluble (or only sparingly soluble) precipitate. This is, indeed, a frequently encountered situation, but it is by no means the only one to give rise to periodic precipitation. For instance, it is well-known that metallic gold deposits can be formed in (what must obviously be) a very different way, e.g. see Hatschek (1911), Holmes (1926), and Kratochvil et al. (1968). Thus, gold chloride in solution can be reduced by a diffusion influx of oxalic acid to yield spaced precipitates consisting of triangular and hexagonal microcrystals (Henisch, 1984). Another example is the precipitation of silver halides by decomplexing, e.g. see Blank and Brenner (1971), Nickl and Henisch (1969), Halberstadt (1967), and Suri and Henisch (1971). Yet another is the deposition of substances from solution by allowing a "reagent" to diffuse which does not actually "react", but serves to change (reduce) the solubility. Thus, for instance, triglycene sulfate, though highly soluble in water, is much less soluble in alcohols, and can therefore be precipitated by adding alcohol to an aquaeous solution. The same basic method has been used by Glocker and Soest (1969) for the deposition of ammonium phosphate, and by Joshi and Antony (1980) for potassium dihydrogen phosphate (KPD).

Solubility modulation will here serve as an example and exercise in computer modeling. Indeed, such a model can be made by a program very similar to the one so far discussed. One could do so, in the first instance, by retaining most of the established code, giving some of the variables new meanings, and making additional modifications only where they are needed. This procedure is helped by the fact that the underlying diffusion equations are bound to remain unchanged. To what extent such models embody physical realism is another matter, one that remains to be strenuously investigated by comparing laboratory observations and computed results. For the moment, we are concerned only with the fact that a plausible model can be established, one that does indeed lead to periodic precipitation.

84   Periodic Precipitation

For this purpose (only), we shall now assume that [A] represents the monomer. This convenient name is here given to a substance which may or may not be simple, but which can precipitate without undergoing chemical change. [B] now represents the incoming solubility modulator, which could be a substance (e.g. alcohol, as above), but it could alternatively be a "condition", e.g. temperature; the computer does not mind in the least. For plausible starting conditions one might assume that that we have a supersaturated solution of [A], of uniform concentration A_{M0}, present in the diffusion medium, such that $A_{M0} > S_{AT}$, where S_{AT} represents the saturation level. This corresponds to the previously used K_S which, being a product of two reagent concentrations, would not be appropriate here. In turn, and corresponding to the previous K_{SP}, we will now assume that there is a supersaturation limit, S_{SL}. Initially, $S_{SL} > A_{M0} > S_{AT}$, but the incoming [B] is assumed to lower S_{SL}, and to upset this relationship. The question is how, and there we are on uncertain ground. We have almost no information on actual relationships, and will have to improvise creatively! For present purposes, we shall therefore propose that (in Basic)

```
let SSL = SSL0*exp[-SEN.B0(X)/10]
```

where SSSL is the supersaturation limit initially set, SEN a user-adjustable sensitivity factor, assumed to be between 0 and 1. The choice of this or an equivalent expression is actually a critical matter for the outcome. Once it is made, we can proceed very much as in previous Chapters, except that the conditions governing $N(X)$ are now simply (in Basic):

```
If A0(X)>SSL then let N(X) = [A0(X)-SAT]/MPG0
```

Of course, $A_0(X)$ must then be appropriately decremented. Subsequently we have growth whenever $A(X)$ again exceeds S_{AT}, and re-solution when (in Basic) $A_0(X) < S_{AT}$. Thus:

```
    DECG = A0(X) - SAT
```
and
```
    INCS = SAT - A0(X)
```

The schematic provisions for grain size effects, as described in Section 4.1 can remain in place. All the small changes are easily made, and the resulting program can take advantage of many of the same display and printout facilities, e.g. see Fig. 6.1.2 below, and Appendix B..

A typical outcome [in this case for $A_{M0} = 30$, $B_{R0}=45$, $D_A=0.2$, $B_D=0.1$] is summarized in Table 6.1.1. It represents a single deposit, the only one appearing under these conditions. Moreover, it is a moving deposit, moving because it grows on one side, while it re-dissolves on the other. In the course of that, it also becomes thicker, as the entries show. Instances of this kind have, indeed, been observed in the laboratory, e.g. see Kirov (1972) and Hermans (1947).

Table 6.1.1

Position of Moving Monomer Deposit Masses as a Function of Time.

$T^{(1/2)}$ / X	6.0	8.9	10.5	12.5	15.4	16.2	17.9	19.8	21.7
38									3.4
39								4.6	32.3
40							4.1	32.4	34.1
41						9.3	31.5	33.3	32.9
42					26.4	32.5	33.5	32.0	29.1
43				12.4	33.1	32.6	31.2	19.3	
44			11.4	31.7	30.1	24.4	8.3		
45		12.7	30.8	29.6					
46		29.9	21.4						
47	22.3	9.6							
48	10.5								

In Section 4.3, we discussed space-time relationships and, in particular, the proportionality between distance covered by diffusion and the square root of the corresponding times. It is of interest to ask whether that proportionality is still maintained in this case, where, to be sure, diffusion again represents microscopic movement, but where we now have also the macroscopic movement of the deposit as a whole. Fig. 6.1.1 plots the position of the maximum mass (a crude criterion) against the square-root of time, and shows that the proportionality continues to be maintained with satisfactory accuracy. Even so, it is easy enough to show that the agreement can be further and, indeed, sensationally sensationally improved by plotting the "center of gravity" of each deposit, instead of the position of the maximum mass. The famous Random Walk relationship evidently governs the proceedings in profound ways, ways that are sometimes transparent, sometimes not.

86   Periodic Precipitation

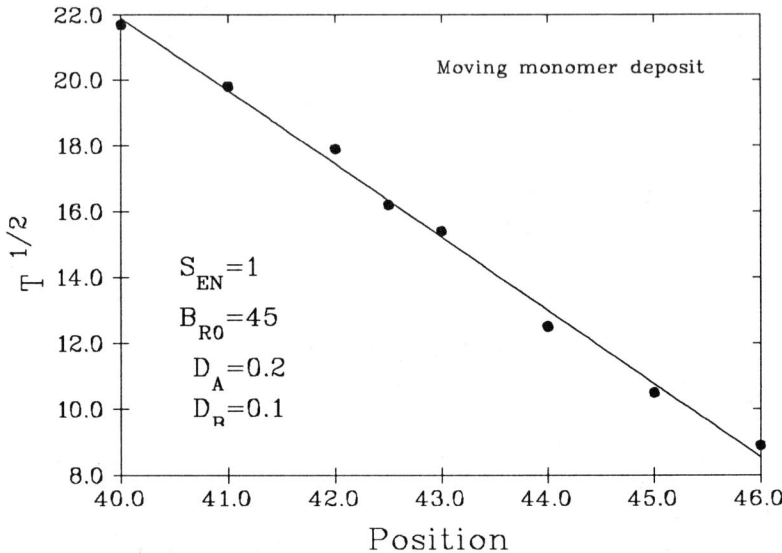

Fig. 6.1.1. Monomer precipitation by solubility modulation. Position of a single moving deposit (based on maximum deposit mass) as a function of $T^{1/2}$; test for linearity. Conditions as for Table 6.1.1.

Fig. 6.1.2 shows banding, in appearance very similar to that previously discussed, but now due to very different causes. No doubt other models can be devised along such lines, modulating S_{AT} and S_{SL}.

6.2 Banding of Particle Distributions by Ostwald Ripening.

Throughout Sections 4.1 to 6.1 above, certain schematic allowances were made for the fact that grains (particles) of different sizes grow and re-dissolve (other things being equal) at different rates; see equations 4.1.7 and 4.1.8. However, one assumption remained untouched, namely the constancy (independence of grain size) of K_S and S_{AT}. This assumption will now be relaxed, not actually to introduce a new order of sophistication into the model, but in order to show that grain size effects can themselves lead to banding, and can do so after the formation of precipitates (or sol particle distributions) as such. Models which describe this type of phenomenon are generally known as Competitive Particle Growth (CPG) models, a term which refers to the fact that larger grains grow at the expense of smaller ones. This means that banding, as such, can result from at least three distinct types of mechanisms: (a) binary diffusion-reaction, as in chapters 1-5, (b) monomer precipitation, e.g. provoked by solubility modulation, as in Section 6.1, and (c) competitive particle growth, also called Ostwald Ripening.

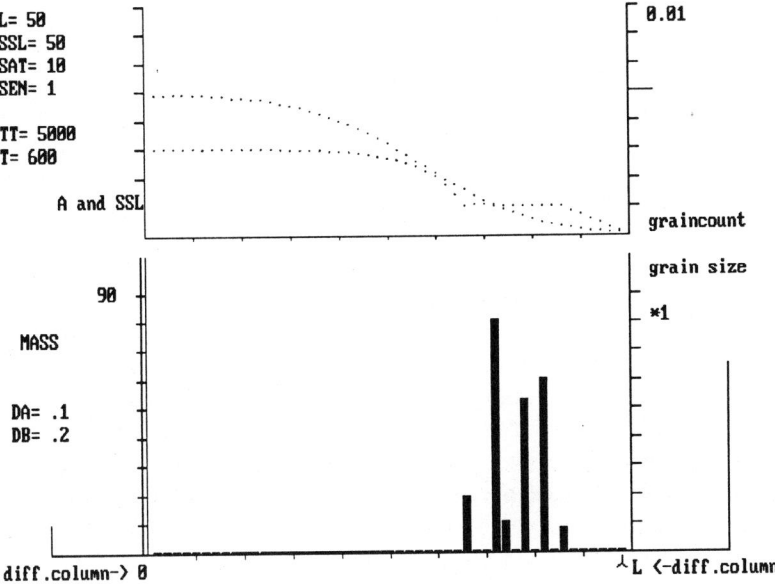

Fig. 6.1.2. Monomer precipitation by solubility modulation. An example of banding [for $B_{R0}=45$, $A_{M0}=30$, $D_A=0.1$, $D_B=0.2$]. The whole band system "moves" to the left in the course of time, with deposits dissolving on the right and re-forming on the left. The deposits here shown at $X=39$, 41 and 43 are all in the process of re-solution.

The notion of competitive particle growth is quite old, e.g. see Ostwald (1925), but the realization that this process can by itself be the cause of banding in varying degree has been explored only in comparatively recent years, e.g. see Ahn and Tien (1976), Lovett et al. (1978), Feeney et al. (1983), Ross (1984), and Field and Burger (1985). Indeed, this mechanism is now frequently invoked as a way of explaining the existence of low density precipitate matter between the prominent deposit layers, e.g. see Kai et al. (1982, 1983). In Section 5.9 above, a similar need was served by running hypothetical experiments "in the probabilistic mode". Of course, both mechanisms may be at work in practice.

The CPG mechanism was originally formulated in response to the observation that, because of surface tension, K_S (or S_{AT}) depend on particle size. Larger particles maintain equilibrium in solutions of lower K_S (or S_{AT}), and it is this fact which allows larger particles to grow at the expense of smaller ones; see also Henisch (1984). By itself, this process would lead only to a coarsening of precipitates (or sols) and that is what the term Ostwald Ripening has always signified. However, when any kind of asymmetry is superimposed on such a system, e.g. in the form of a diffusion gradient, as proposed by Flicker and Ross (1874), then a general ordering transformation can take place which produces banded deposits out of uniformly distributed ones. Computer-generated examples will be found in Sections 6.3 and 6.4. Feinn et al. (1978) describe actual systems of this kind, involving lead iodide.

The basic model envisages a system in which a precipitate (of substance [C]) pre-exists, uniformly distributed throughout the diffusion medium, and consisting of N spherical particles, each of radius R_0 and density R_{HO}, at every location X. [For a harmless, but highly convenient, retreat from this ideal of total uniformity, see Appendix C, where allowance is made for grain-free zones of user-adjustable width near the boundaries.] N is here expressed as a concentration, *not* a numerical count. The particles are in equilibrium with an initial, uniform solute concentration ($C_{E0} + C_{EM}$). The two components of this equlibrium concentration play different roles. It is assumed that the general equilibrium concentration C_E of [C] is a function of the particle radius R, and is given by

$$C_E = C_{E0}.R_0/R(X) + C_{EM}. \qquad (6.2.1)$$

C_E thus diminishes with increasing radius, but it can never be smaller than a certain minimum, namely C_{EM}; (Fig. 6.2.1). The limit C_{EM} is thus the equilibrium concentration of infinitely large particles. (A safety clause which prevents $R(X)$ from falling to zero is very necessary, and can easily be included in the code.) Whether all the equilibrium situations suggested by equation 6.2.1 can actually be reached in the system for an arbitrarily chosen value of $R(X)$ is another matter. It depends, of course, on whether enough material is available to transact the zero-sum interplay between particles and solution. Thus, particles of very small radius would demand a very high value of C_E for equilibrium, but it may not in practice be possible to achieve such high concentrations, in which case the particles will gradually disappear. Until they do, they will co-exist with the solution, albeit in non-equilibrium. In any event, it should be understood that the specific form of equation 6.2.1 is not obligatory; all that is needed is for C_E to be a plausible function of R, and functions more plausible than the purely schematic equation 6.2.1 may indeed be found.

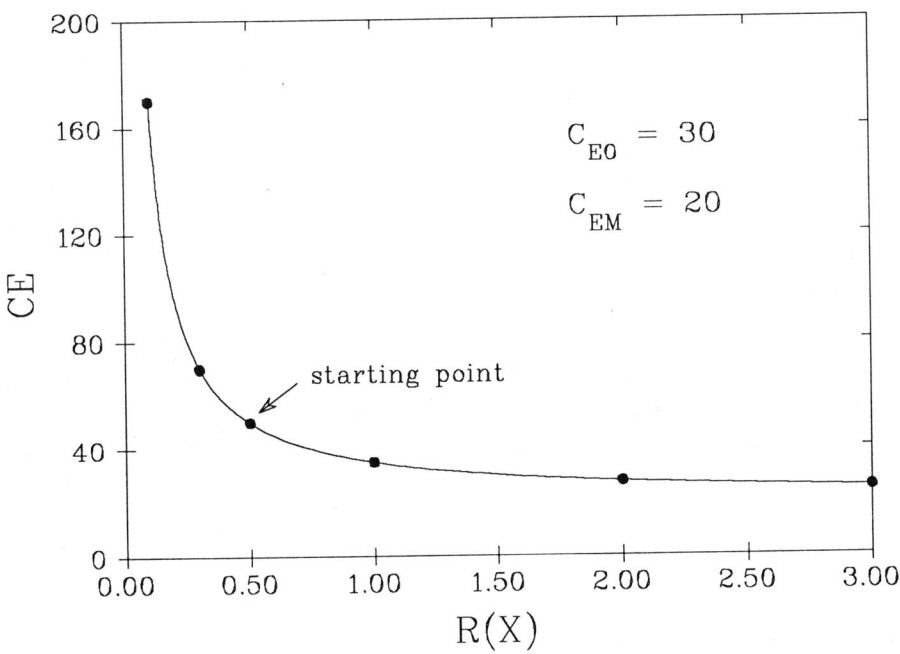

Fig. 6.2.1. Equilibrium concentration of solute as a function of co-existing particle radius, based on the (hypothetical) equation 6.2.1.

The rate of particle growth at any location X is taken to be directly proportional to $C(X)-C_E$:

$$dR/dT = K.[C(X)-C_E] \qquad (6.2.2)$$

where K is an adjustable rate constant. The number (N) of particles remains unchanged, but wherever there is particle growth or shrinkage, the solute concentration has to be appropriately adjusted, that is to say, in accordance with the change of grain mass, by $N.R_{HO}.4.\pi.R^2.dR$ in unit time. In terms of Basic, we thus have, by analogy with equation 1.4.6:

$$C(X,T+1) = D_C \cdot C(X-1,T) + (1- 2D_C) \cdot C(X,T) +$$
$$D_C \cdot C(X+1,T) -$$
$$N \cdot R_{HO} \cdot 4\pi \cdot [R(X,T)^2] \cdot K \cdot [C(X,T) -$$
$$C_{E0} \cdot (R_0/R(X,T)) - C_{EM}] \qquad (6.2.3)$$

This is actually the core of the program. Equation 6.2.3 is iteratively solved in the presence of a small concentration gradient, established by making the "high" reservoir concentration C_{RH} (at $X=0$) slightly greater than $C_{E0}+C_{EM}$, and the "low" reservoir concentration C_{RL} (at $X=L$) slightly smaller.

Of course, such a simple formulation can't hold indefinitely; it grinds to an ignominious halt, for instance, when particles have become so large that the next solution decrement (equ. 6.2.3) makes the concentration negative! One of the problems is that we have here assumed every particle-solution transaction to be instantaneous. In practice, this cannot be, and a further level of sophistication would be needed to build the micro-dynamics of particle growth into the model (Task reserved for the second edition of this book, if there should ever be one).

6.3 Experiments with Competitive Particle Growth; a Typical Run.

All the equations of Chapters 1-5 are deterministic (when run with D_{NC} set to unity), and if the outcome is often intuitively unpredictable (as, indeed, it is), this is due to the fact that many interacting parameters are involved. In so many ways, the (equally deterministic) model described in Section 6.2 simpler, and its transactions are in the hands of far fewer adjustable parameters. It is therefore all the more intriguing to note that its outcome is, if anything, less predictable. In this case, the unpredictable behavior has little to do with the number of user-determined parameters, and everything with the inherent character of non-linear mathematics. We operate here on the borders of two new and highly topical fields: "catastrophe theory" and "order-out-of-chaos theories". Beginning with a uniform distribution of particles, we end up with bands (when we do!), bands thin, thick, closely spaced, widely spaced, consisting of light particles or heavy ones, or of no particles at all, and there is no immediate way of forecasting the consequences of even (apparently trivial) parameter adjustments. Typical experiments are described below and in the next Section.

We will begin with an episodic account of a run made with a set of the default parameters which have been found to be convenient for this purpose. The results are shown in Figs. 6.3.1 a-d. In Fig. 6.3.1a we see that the initial mass distribution is practically uniform, except in the immediate vicinities of the boundaries, which have been kept free of preformed particles. It is simple enough to color-code the video-display (see Appendix C), so that growing, stable, and dissolving masses appear in distinctive colors. Such a display (but, alas, not here in Fig. 6.3.1a) shows that, at a time $T=100$ for instance, the particles on the left are growing, those on the right are shrinking, and those in-between ($X=13$ to 37) are stable, at any rate in the sense that the are changing only trivially at this stage, if at all. As far as it goes, and just for once, this is entirely in accordance with expectations.

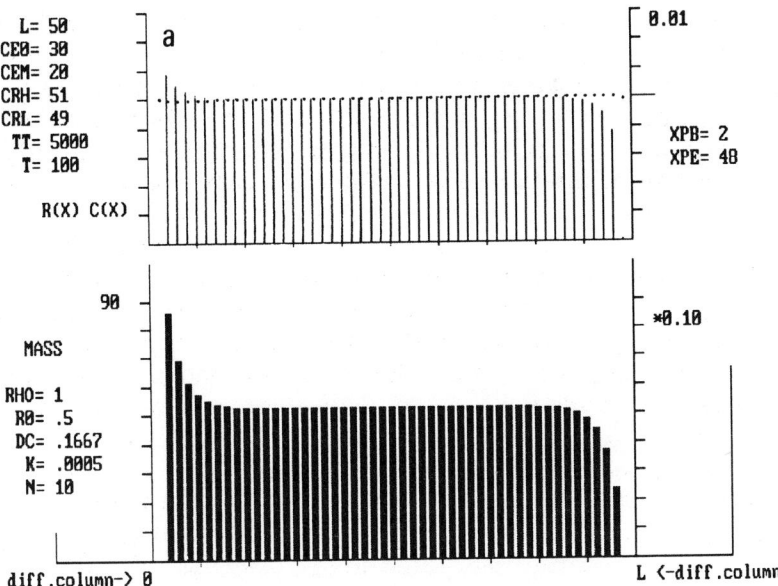

Fig. 6.3.1 Experiments with competitive particle growth; typical run with a set of default parameters. (a) $T=100$, (b) $T=300$, (c) $T=600$, (d) $T=1045$.

The same system looks significantly different at $T=300$ (Fig. 6.3.1b), even if we ignore the change of scale. Banding is in progress and, within the $X = 5$ to 10 region, the particles have already disappeared. Were our Figure in color, it would show a growth region for $X = 35$ to 43. Throughout the observations, it is a common sight to find growing deposits next-door to shrinking ones (which nourish them).

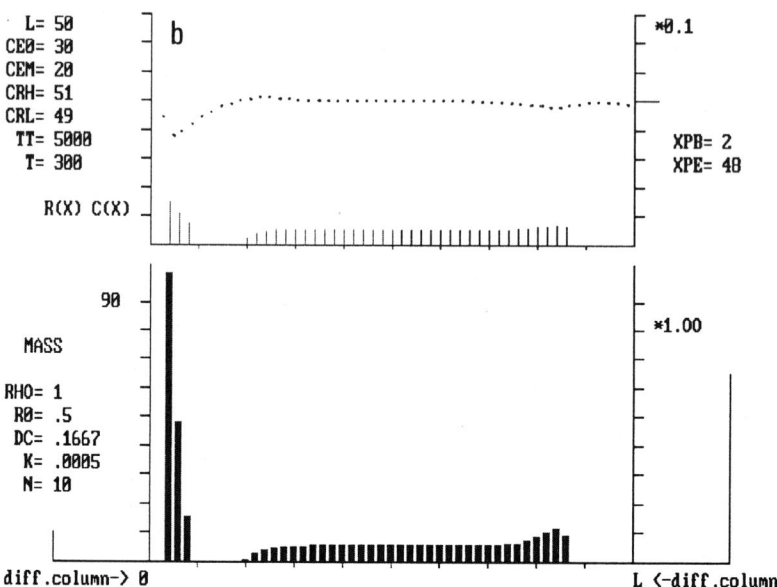
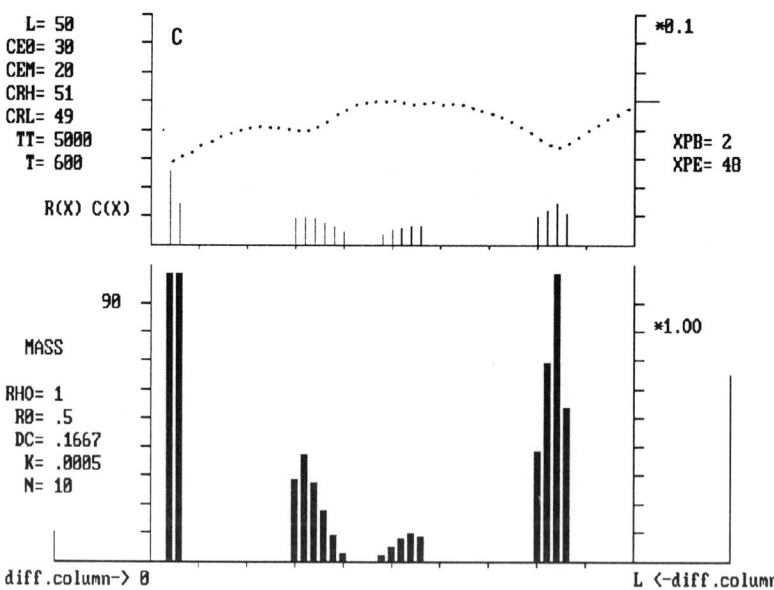

Fig. 6.3.1 (b) above, $T=300$, and (c) below, $T=600$.

By the time $T=600$ (Fig. 6.3.1c), we have very pronounced banding. The $C(X)$ contour reflects the fact that larger particles tend to co-exist with lower solute concentrations: every deposit cluster is associated with a corresponding dip.

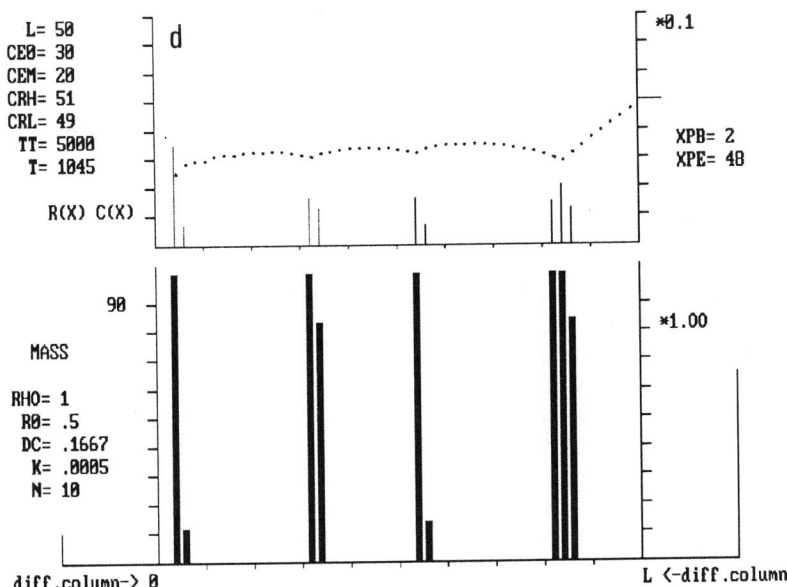

Fig. 6.3.1 (d) $T=1045$.

By the time $T=1045$ (Fig. 6.3.1d), the originally uniform distribution has been transformed into a highly structured pattern, consisting of only four thin deposits. Note that the magnitudes shown are no longer significant, since the demands exceed the available display space. Of course, the actual values of $M(X)$ can be obtained from the numerical readout.

At this stage, all the deposits are growing, but at $T=1700$ the particle cluster at $X=16$ oscillates betwen growth and stability. At $T=1915$ it becomes quasi-stable for a while until, at $T=2330$, it begins to dissolve. The deposit at $X=27$ likewise has a complex career. After such a long time, the $C(X)$ contour shows large gradients at the two boundaries, and those ensure that the deposits at $X=2$ and $X=42$ will continue to grow for a long time. In the central region,, $C(X)$ is much lower than it originally; a good deal of the [C] originally in solution has been used up for particle growth, and the new particles, in turn, can co-exist with a lower solute concentration.

One might ask why the deposits are not symmetrical, considering that the descriptive parameters in this case are symmetrical; they are, indeed, but equation 6.2.1 is not.

Such a run illustrates the type of observation that can be readily made, and it also promotes confidence in the general coherence of the method and its results.

6.4 Experiments with Competitive Particle Growth; Effect of Parameter Variations.

The program can now be used to show how the user-determined system parameters affect the outcome. As always, we will conduct this exploration by varying one parameter at a time, for comparison with the results obtained in Section 6.3. For the run described by Fig. 6.3.1, the number of particles at each site X was fixed at 10. How important is this choice? Comparative test runs with $N=1$ and $N=30$ show how the system reacts to the changed starting conditions.

When $N=1$, and the particles have the same original size, the particle-solution interaction is greatly reduced, and cannot play much of a role; particles may, for instance, dissolve, but this does not contribute an appreciable amount of free solute. Whatever drama there is, comes later, when the particles which grow have in fact grown substantially. At $T=400$ (Fig. 6.4.1a) we see such a build-up at small values of X, and everything seems set for a continued narrowing of that single band. However, we are in the wrong field for confident predictions; the expectation is not fulfilled. On the contrary, shrinkage gives way to buildup, and the resulting situation for $T=645$ is shown in Fig. 6.4.1b. Not for long, though; at $T=830$, a middle region ($X=8$ to 10) of that band actually begins to dissolve, spreading rapidly, while mass (indeed, its mass) continues to be accumulated at its borders. Thus, by $T=1500$, the picture looks quite differently (Fig. 6.4.1c). In due course, the lower density intermediate material, disappears, leaving only two isolated bands, consisting of grains very much larger than those at the start (Fig. 6.4.1d), "coarsening", indeed. With ongoing time, those two bands come to be more and more sharply defined; both eventually become single lines (at $X=2$ and $X=20$).

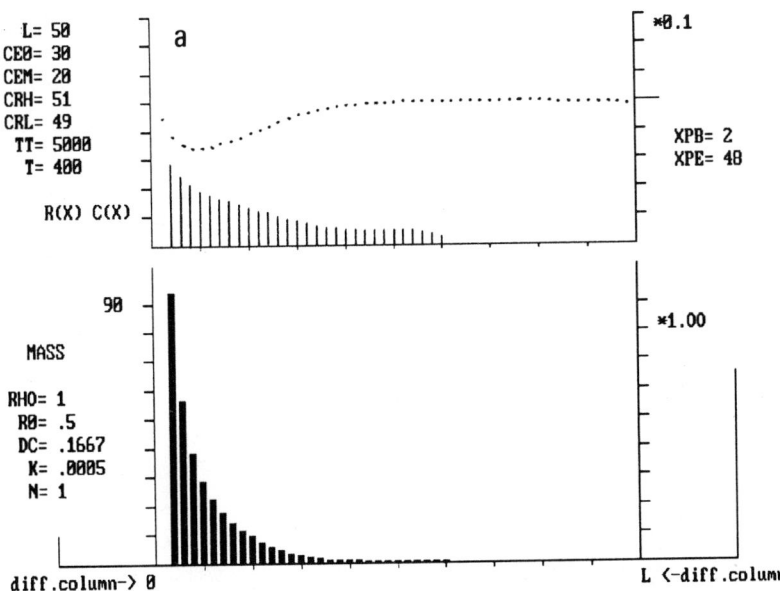

Fig. 6.4.1. Experiments with competitive particle growth; effect of parameter variations, for comparison with Fig. 6.3.1. Initial number of particles equivalent to a concentration $N=1$. (a) $T=400$, (b) $T=645$, (c) $T=1500$, (d) $T=2500$.

In contrast, when the original number of particles is high, events proceed much more rapidly. This can be seen, for instance, by comparing the above cases with a situation report for $N=50$ at $T=1215$; Fig. 6.4.2. Note the $C(X)$ contour, with its upward bulge in the low density region.

The next experiment will contrast two situations, controlled by different values of C_{E0} and C_{EM} chosen, however, in such a way that the starting concentration, $C_{E0}+C_{EM}$ remains the same. The new results in Fig. 6.4.3 are thus be directly comparable with those shown in Fig. 6.3.1c. The pattern is not profoundly different, but note that the $C(X)$ contour is more modulated now, as we would expect from equation 6.2.1.

All the above parameter changes have been concerned with the physical setup; the next change concerns an algorithm parameter, namely the growth coefficient K (see equation 6.2.2). When K is diminished, we know simply that less and less is going to happen. For a small increase,

96   Periodic Precipitation

to $K=0.0006$, the situation is not radically changed, but for $K=0.001$ implying more rapid growth, the banding is now much more pronounced (Fig. 6.4.4); Compare Fig. 6.3.1b. For once, the is what we would expect. The situation shown is not stable, of course; it develops eventually into five isolated bands.

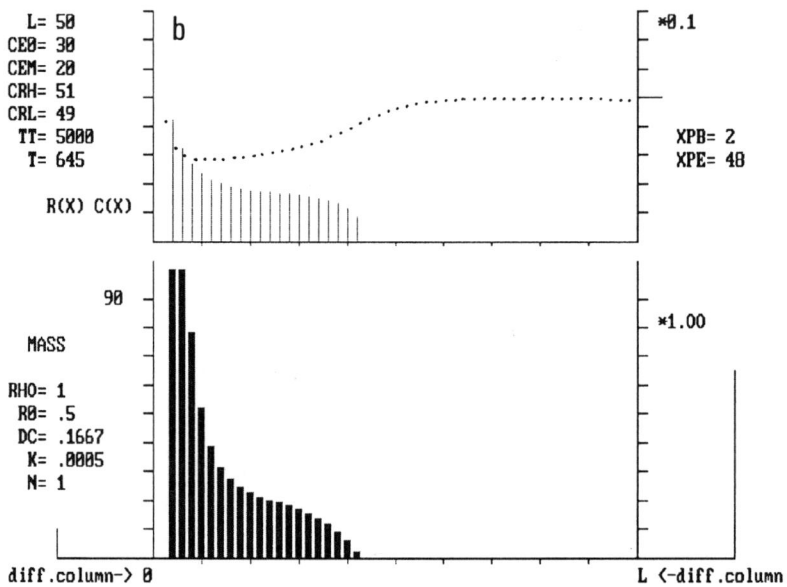

Fig. 6.4.1 (b) $T=645$.

The sensitivity of the system to changes of initial particle radius R_0 is shown in Fig. 6.4.5a, which envisages substantially larger grains. This presentation is for all practical purposes comparable with Fig. 6.3.1b. Note the strong modulation of the $C(X)$ contour, brought about by the fact that a great deal of material is now involved in the solution-particle interaction. By the time $T=1045$ (Fig. 6.4.5b) and, in due course, $T=2500$ (Fig. 6.4.5c) we see just how drastic (and unpredictable) the consequences of doubling the radius have been. In computer work, it seems, we manage to be equally pleased when predictions are confirmed (strengthening our pride in the algorithm), as we are when totally unforeseen results are recorded (satisfying our ambitions as pioneers); few other endeavors offer such privileges.

Experimentation with Monomer Systems

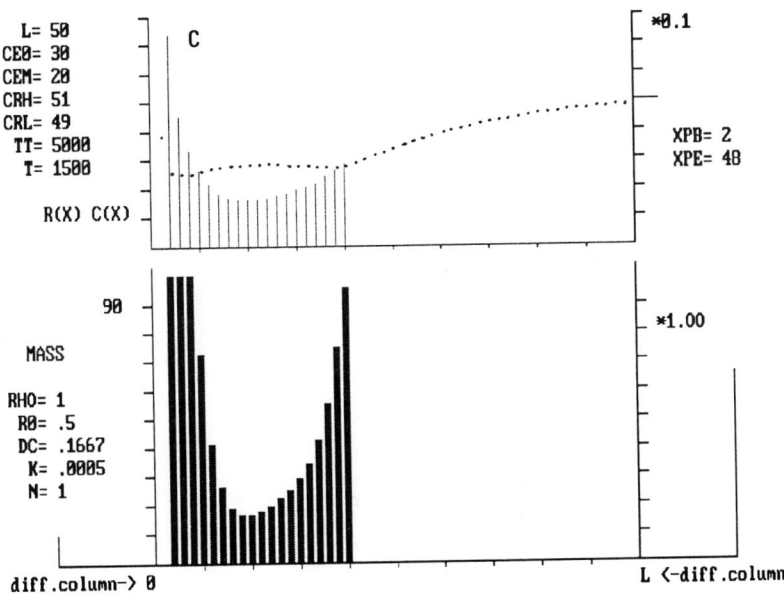

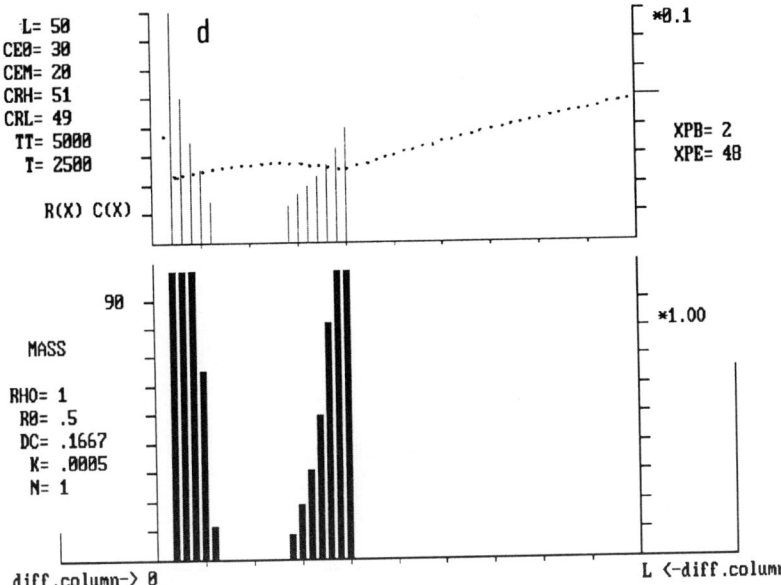

Fig. 6.4.1 (c) above, $T=1500$, (d) below, $T=2500$.

Periodic Precipitation

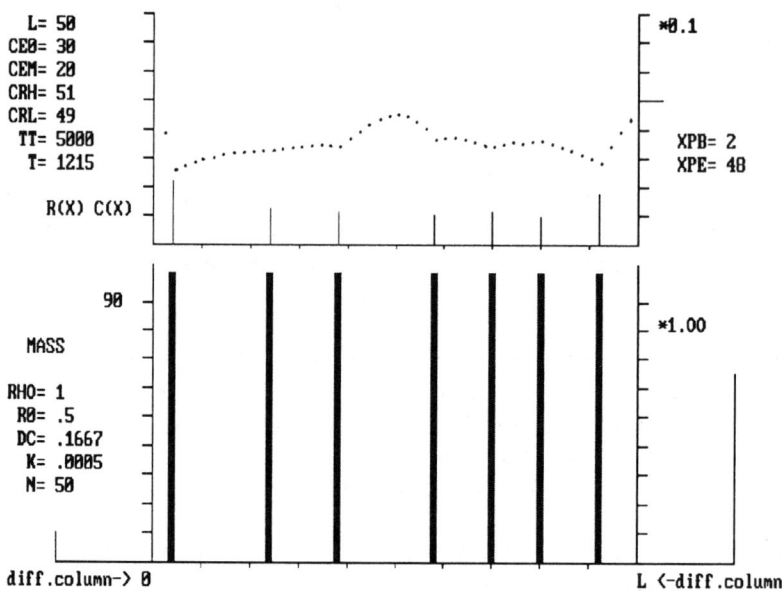

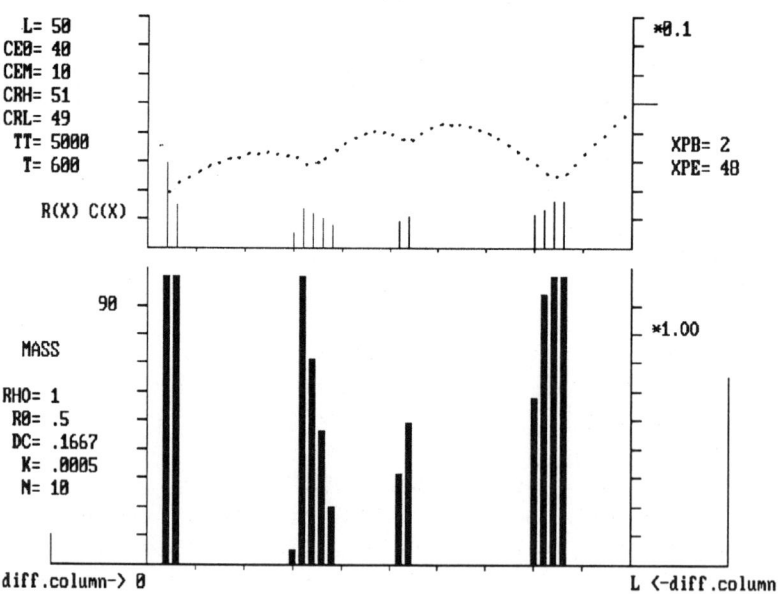

Fig. 6.4.2 (above); Fig. 6.4.3 (below)

Fig. 6.4.2. Experiments with competitive particle growth; effect of parameter variations, for comparison with Fig. 6.3.1. Initial number of particles equivalent to a concentration $N=50$. $T=1215$.

Fig. 6.4.3. Experiments with competitive particle growth; effect of parameter variations, for comparison with Fig. 6.3.1. Effect of particle-solution equilibrium conditions. $C_{E0}=40$, $C_{EM}=10$, $T=600$.

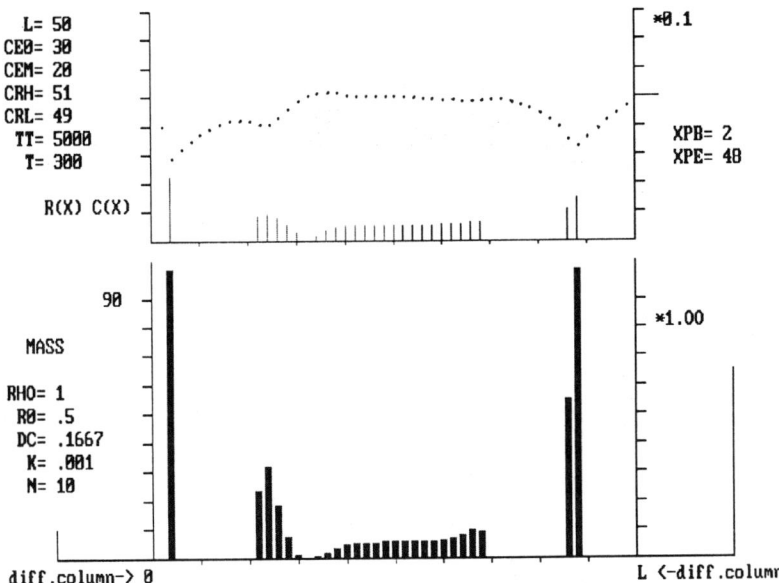

Fig. 6.4.4. Experiments with competitive particle growth; effect of parameter variations, for comparison with Fig. 6.3.1. Growth coefficient, $K=0.001$. $T=300$.

The present survey of results makes no claim of being exhaustive; its purpose is, above all else, to illustrate a research opportunity, while the joys and excitements of further experimentation are left (as Victorians would have put it) to the gentle reader.

100 Periodic Precipitation

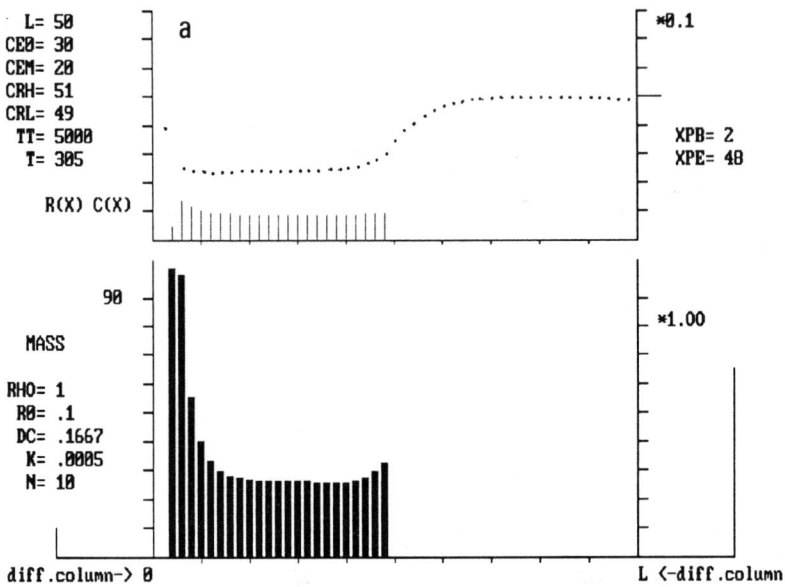

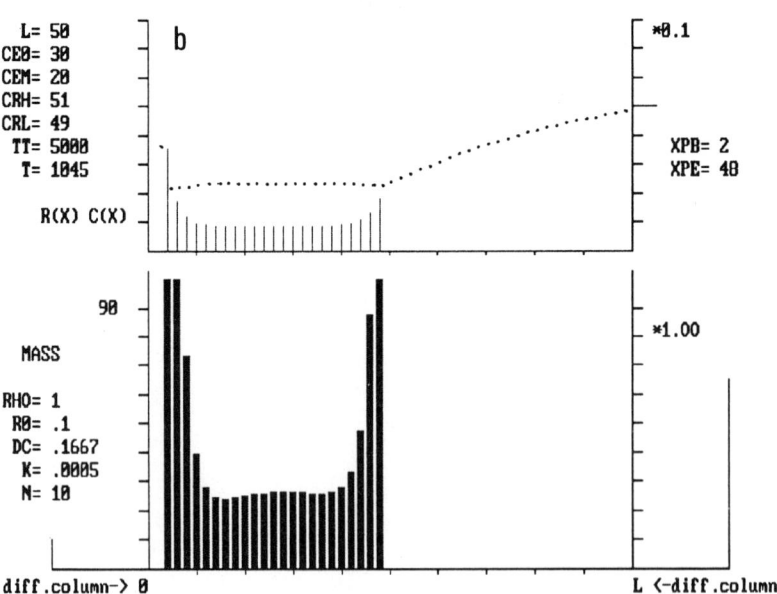

Fig. 6.4.5 (a) above, (b) below.

Fig. 6.4.5. Experiments with competitive particle growth; effect of parameter variations, for comparison with Fig. 6.3.1. Initial particle radius, $R_0=0.1$. (a) $T=305$, (b) $T=1045$, (c) $T=2500$.

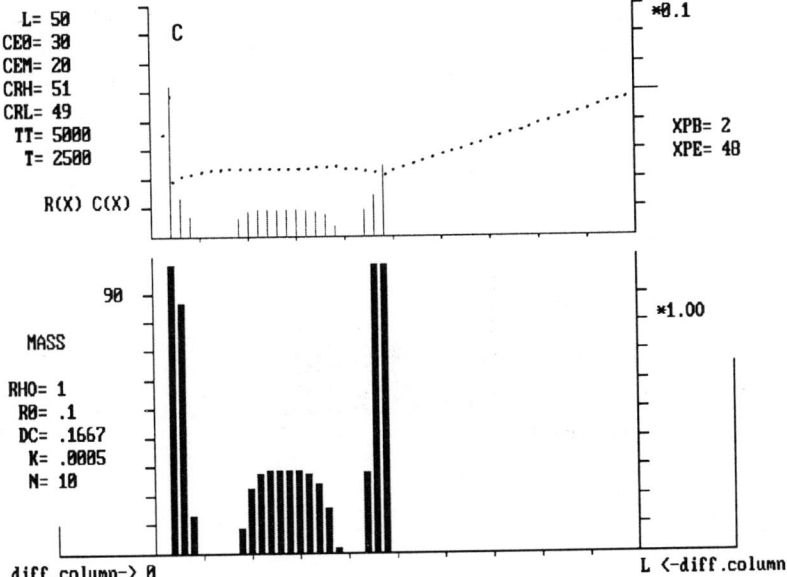

Fig. 6.4.5 (c).

APPENDIX A

Precipitation by Binary Reaction; Practical Software Implementation [Programs PPBIN].**

The coding fragments discussed in Chapters 1-6 should certainly help the reader to devise special-purpose programs, but they do not by themselves amount to a comprehensive software package with a viable "user-interface". Such a package is available for use in conjunction with this book (see below). For those who prefer to create their own program "shell" a strategic plan can easily be outlined. It takes advantage (again) of the highly structured nature of the True BASIC language and its convenient built-in graphics facilities. However, any modern language will serve in principle. Moreover, there is nothing compelling about the plan, programing being an art as much as a science. Still, most of the features outlined have to be included in one way or another, and a listing of typical provisions will be a help in clarifying the logical connections.

A comprehensive program package would consist of four types of features: (1) set-up routines which are called in order to establish the running conditions, and before any interations begin, (2) calls to special-task subroutines which are at the core of algorithm and its iteration procedure, (3) key input intercepts which permit the user to play an interactive role while the program is running, and (4) calls to subroutines which have to do with the repeated running of the program in order to explore the probabilistic mode. In the first group, one would include the following subroutines or equivalents:

SUB TITELPAGE: Self-explanatory opening screen, which tells the user about the special pleasures and opportunities in store. It also offers an opportunity for calling

SUB HELP, before proceeding. On the Help screen, the various switches are explained which select the display modes. "Press RR to continue".

SUB PARAMSETUP: Assignment of the default values to the various parameters.

SUB SHOWPARAM1: List of parameters concerned with the definition of the physical setup envisaged by the computer experimentation. The display for "SHOWPARAM1", but typical of all "SHOWPARAM" subroutines, is shown in Fig. A1. It lists the default values, and prompts for user input. An answer of Y or y to ANY CHANGE? leads to a further prompt, asking the user to specify the variable to be changed, and to give its new value. At some stage, the user will accept the list, and this will activate a direct call to SUB VALIDATE1, containing error-trapping routines which ensure that the parameter values user-selected in SHOWPARAM1 are within acceptable limits. If they do not, the user is returned to SHOWPARAM1. Such a return, by its very nature unexpected, should be taken as a sign of the fact that "something is wrong" with the selected pattern of values on that screen. Inspection will quickly reveal what the problem is.

```
            PARAMETERS OF THE PHYSICAL SETUP GROUP

   L   =  50      LENGTH OF THE DIFFUSION SYSTEM (25=<L>=400)

   AR0 = 100      INITIAL RESERVOIR CONC. OF (A).  AR0=<100

   BR0 = 100      INITIAL RESERVOIR CONC. OF (B).  BR0=<100

   AM0 =  10      INITIAL CONC. OF (A) IN DIFFUSION MEDIUM. AM0<100

   XAM =  49      PRE-CHARGING DISTANCE FROM (A) RESERVOIR. 0<XAM<L

   RDC = .006     RESERVOIR DEPLETION COEFFICIENT. RDC<=0.1

   SSA =   0      SYSTEM STATUS AT (A)-END. (CLOSED=0, OPEN=1)

   E   =   0      ELECTRIC FIELD. NORMALLY (-1)<E(+1)


                     ANY CHANGE?  Y/N
```

Fig. A1. Parameter display (Physical Setup Group) of Program PPCPG01. Default values and provision for changes. Display format typical of all "SHOWPARAM" subroutines.

SUB SHOWPARAM2: List of parameters concerned with the basic algorithm, namely (in Basic):

Appendix A   105

```
KS   ! the solubility product,
KSP  ! the precipitation product,
ER   ! the "equality range", (stoichiometry)
MPGO ! the initial mass per grain,
DNC  ! the "deterministic nucleation
       coefficient",
PHF  ! factor governing interaction with [H]
DA, DB, DH ! the diffusion constants of
         [A], [B], and [H].
```

Display of the default values, and provision for user-determined changes. This would, in turn, call SUB VALIDATE2, containing error trapping routines which ensure that the parameter values user-selected in SHOWPARAM2 are within acceptable limits. If not, return to SHOWPARAM2. Above and elsewhere: ! = REM in other dialects of BASIC.

SUB SHOWPARAM3: List of parameters concerned with the running of the program, namely (in Basic):

```
TT  ! the total  time of each run,
TRD ! the video updating interval,
SPT ! the special printout time,
RT  ! the total number of runs,
HIC ! a program choice relating to the
      interaction with [H]; see Section 5.6,
NUC ! a program choice relating to the
      implementation of the random element;
          see Section 5.9.
```

This subroutine, in turn, calls SUB VALIDATE3, containing error trapping routines which ensure that the parameter values user-selected in SHOWPARAM3 are within acceptable limits. The default value of S_{PT} may be set above T_T, in order to avoid an automatic and unwanted printout. Indeed, this is always done when operating in the probabilistic mode ($D_{NC}<1$ and $R_T>1$), since no individual printout would be significant.

A word about the choice of T_{RD}. The default value is 5, calling for a video update whenever T is divisible by 5. Making T_{RD} greater speeds up the proceedings, but involves the risk of mnissing something interesting on the monitor. Conversely,

$T_{RD}=1$ would catch every interesting event, but would make the program very slow.

SUB SHOWPARAMF: List of parameters concerned with the management of the electric field regime, to be called only when the electric field is not set at zero:

```
FF   ! governing whether the field
        should be static, temporary or
     alternating,
PRI  ! phase reversal index,
     for alternating fields,
THP  ! frequency of the alternating field
     (more precisely, the time-value of the
     half-period),
FAT  ! field application time
     (for temporary field),
FRT  ! field removal time,
     (for temporary field).
```

This subroutine, in turn, calls SUB VALIDATEF, containing error trapping routines which ensure that the parameter values user-selected in SHOWPARAMF are within acceptable limits.

SUB SYSTEMSETUP: Subroutine which serves to dimension and initialize the array variables, in response to the user-choices made above. Here also, the initial reagent concentrations in the medium are set, and the boundary conditions functionally imposed.

SUB DISPLAYSETUP: Subroutine which establishes the graphics display on the monitor (axes, scales, numerical information, status reports, etc.)

All these tasks have to be completed first. Once this is done, one proceeds as outlined in Section 4.4:

```
from J=1 to RT
   from T=1 to TT
      ...................! main program
   next T
next J
```

where T is the time counter, and J counts the number of runs. Subroutines called for each value ot T:

SUB RECALC: Implementation of the diffusion algorithm.

SUB GRAINCOUNT: Routine to establish the value of $N(X)$. -

SUB GRAINGROWTH: Governing mass transfer to the grains.

SUB RE_SOLUTION: Governing mass transfer from the grains.

When the programs are run in the probabilistic mode ($D_{NC}<1$, $R_T>1$), one more provision is needed during each run, namely:

SUB TIMEDISPLAY, which tells the user what value of J and T the programn has reached.

Two more subroutines are needed after each run of length T_T, namely:

SUB ACCUMULATE: Cummulative placement of the results (for each value of X) after each run into appropriate array variables.

SUB RESET: Provision for ensuring that each run begins with the same starting conditions, un-influenced by previous runs.

Then one proceeds with:

SUB AVERAGE: Calculation of the average results after $J=R_T$ runs.

SUB FINALSHOW1: A graphics display which concerns itself with the presentation of the averaged information in the form of simple graphs. It permits a call to:

SUB FINALSHOW2: Provision for the display of the final (averaged) information in the form of a pictorial simulation, or else in the form of a hard-copy printout, or both.

Throughout the operation, key input routines are interrogated after each iteration T, and allow the user to select:

SUB CDISPLAY, upon pressing C at any time,

SUB MDISPLAY, upon pressing M at any time. On the M-display, growing deposits are shown in dark red, deposits in the process of re-solution (however slowly) in bright red, and stable deposits in yellow. However, the term "stable calls for a working definition, since a demand for total stability would be difficult to satisfy! We shall here call "stable" any deposit which has not changed by more than 1 part in 10⁴ since the last pass. There is also a:

SUB GDISPLAY, reached by pressing G at any time, which adapts the general DISPLAYSETUP in the manner described in Section 5.1.

Alternatively, the key input routines can call for:

SUB NUMDISPLAY, by pressing N at any time. This subroutine stops the clock and displays an overview of the existing status of events in numerical form. The data immediately on the screen cover 16 mid-range values of X, which may or may not be the values in which the user is immediately interested. All other values can be brought into view by pressing PgUp or PgDn.

As an alternative to visual readout, there is provision for hard copy, via:

SUB PRINTREPORT, reached by pressing P at any time. The first stage is to offer the user a printout of the parameter choices, and for this purpose the program requests the X-values for which printed data are desired. This avoids a good many unnecessary printout operations, e.g. in X-ranges over which all critical values are simply zero. Upon all subsequent calls to this subroutine, the initial parameter printout is omitted.

None of these implementations is actually complicated but, without question, the task of coding is often tedious. Fortunately, the complete program is commercially available. In compiled form, it can be used on its own, with full control over all the user choices, and without any language requirements. In source code form, it can be run only from within the True Basic language environment, which gives the user control not only over the variables, but over the program structure itself.

APPENDIX B

Monomer Precipitation by Solubility Modulation; Practical Software Implementation [Programs PPMON**].

Most of the above comments apply also to this package, except for the changes outlined in Section 6.1. In the practical implementation that is commercially available, a number of matters have been simplified by the omission of provisions relating to electric field effects, waste product interaction, and probabilistic mode operation. The C-Display is not really needed, since the corresponding information can be included in the (now vacant) upper half of the M-Display. In all other respects, the implementation is the same.

SUB SHOWPARAM1 handles the same values as before, except E.

SUB SHOWPARAM2 now displays the variables appropriate for the algorithm described in Section 6.1, and offers the opportunity for user-determined changes. The variables are (in Basic):

```
SAT   ! saturation level,
SSL0  ! initial value of the supersaturation
        level,
MPG0  ! initial number of particles per grain
SEN   ! index (=<1) which govers the
        sensitivity of the supersaturation level
        to [B]
DA, DB ! diffusion constants of [A] and [B].
```

SUB SHOWPARAM3 concerns itself with:

```
TT   ! the total time of each run,
TRD  ! the video updating interval, and
SPT  ! the special printout time.
```

There is no provision for a probabilistic mode, and therefore no need for automatically scheduled multiple runs. The G-Display is available as before (press G at any time), and so are all the numerical (press N) and hard-copy (press P) readout facilities.

APPENDIX C

Monomer Banding by Competitive Particle Growth; Practical Software Implementation [Programs PPCPG].**

The program makes use of the same basic "shell" described in Appendix A, but of the display modes only the

SUB MDISPLAY is implemented. It plots mass in the lower half (growing in dark red, quasi-stable in yellow, and re-dissolving in bright red), and particle radius (green), as well solute concentration (green) in the upper.

The extreme mass ratios here encountered make the routine scaling provisions difficult to operate; displays have a habit of "collapsing" when there are still interesting things to see. Some adjustments have been made in response to this need. Thus, a mass bar which should, in principle, exceed the available display space is at times truncated to the maximum height, and thus accommodated within the picture. The correct numerical information is, of course, available at any time via the numerical readout ("press N") and printout ("press P").

The problem is mitigated somewhat by the fact that $R(X)$ and $M(X)$ contain basically the same information, but do not scale in the same way. Thus, processes which may be difficult to follow in detail in the lower half of the display may be reflected with clarity in the upper half.

Another (user-controlled) way of reducing the scaling problem is to make the "high" reservoir concentration C_{RH} only slightly greater than $C_{E0}+C_{EM}$, or even equal to it. This tends to avoid an otherwise dominant growth at $X=1$, which often exerts a controlling influence over the scale of the display as a whole.

Lastly, it is possible to envisage a system in which the diffusion medium is uniformly filled with particles to a concentration N at each value of X, except for certain marginal zones near $X=0$ and $X=L$. Indeed, provision has been made for the user to select a beginning position (X_{PB}) and an end position (X_{PE}) for that particle distribution. This provision has been implemented in

112   Periodic Precipitation

SUB SYSTEMSETUP, which, in other respects, serves all the functions described before. Over the X-interval filled with particles, we have as the initial state, $R(X)=R_0$. Over the X-interval(s) not so filled, we put $R(X)=R_0/100$, another way of saying "practically no mass here". The distribution limits may be user-imposed symmetrically or asymmetrically.

Of course, the standard input routines are needed, as before, but

SUB SHOWPARAM1 now concerns itself with

```
L   ! the length of the diffusion system
CRH ! the "high" reservoir concentration
CRL ! the "low" reservoir concentration
CEM ! minimum equilibrium concentration
CE0 ! incremental equilibrium conc. for
      R=R0,
XPB ! X-value at which the particle
      distribution begins, and
XPE ! the X-value at which the particle
      distribution ends.
```

The last two variables are included under the heading of Physical Setup, because the sum $C_{E0}+C_{EM}$ determines the starting concentration of [C].

SUB SHOWPARAM2 displays and sets user-determined values of N, the number of pre-formed particles at every location, using

```
RHO ! the particle density
RO  ! the initial particle radius
K   ! a rate constant governing
      particle growth, and
DC  ! the diffusion constant of [C].
```

The correspondiong validation procedures (VALIDATE2) under this heading have been relaxed in order to allow for a wide (albeit cautious!) exploration of strange territories. As so often, enthusiastic scepticism is the best operating mode.

SUB SHOWPARAM3 deals, as before, with T_T, T_{RD}, and S_{PT}, which have the same meanings as in Appendix A. The default value of

S_{PT} can be set (default) above T_T in order to avoid an automatic and unwanted hard-copy printout.

The subroutines RECALC, GRAINCOUNT, GRAINGROWTH and RE_SOLUTION described in Appendix A are here replaced by a single code-passage, called

SUB CPG. This transacts the calculations governing the diffusion of the monomer, now called [C] to distinguish it from "reagents" [A] and [B] previously used. It also calculates new particle sizes, as described in Section 6.2, and makes provisions for the corresponding solute decrements and increments.

There is no provision for a probabilistic mode, nor for reservoir depletion; in this program the reservoirs are regarded as inexhaustible.

REFERENCES

[The S-number(s) after each entry refer to the section(s) in which the work is cited.]

Adair, J., Touse, S.A., and Melling, P.J. (1987) *American Ceramic Society Bulletin* 66, 1490.   {S 1.1}

Ahn, T.M. and Tien, J.K. (1976) *J. Chem. Solids* 37, 711   {S 6.2}

Bajpai, A.C., Calus, I.M., and Fairley, J.A. (1977) *Numerical Methods for Engineers and Scientists.* John Wiley and Sons, London.   {S 1.4}

Bennett, Jr., W.R. *Scientific and Engineering Problem-solving with the Computer.* Prentice Hall, Englewood Cliffs, N.J.   {S 1.1}

Blackman, J. (1987) *Physics Bulletin* 38, 385.   {S 1.1}

Blank, Z. and Brenner, W. (1971) *J. Crystal Growth* 11, 258.   {S 6.1}

Braterman, P. (1989) University of North Texas. Personal communication.   {S 5.3}

Christomanos, A. (1950). *Nature* N4189, 238.   {S 5.8}

Crank, J. (1956) *Mathematics of Diffusion.* Oxford University Press.   {S 1.2}

Dhar, N.R. and Chatterji, A.C. (1925) *Kolloid Z.* 37, 2   {S 5.8}

Feeney, R., Schmidt, S.L., Strickholm, P., Chadam, J., and Ortoleva, P. (1983) *J.Chem.Phys.* 78, 1293.   {S 6.2}

Feinn, D., Ortoleva, P., Scalf, W., Schmidt, S., and Wolff, M. (1978) *J.Chem.Phys.* 69, 27.   {S.6.2}

Field, R.J. and Burger, M. (1985) *Oscillations and Traveling Waves in Chemical Systems.* J. Wiley, New York.   {S 6.2}

Flicker, M. and Ross, J. (1874) *J.Chem.Phys.* 60, 3458.   {S 4.3}

García-Ruiz, J-M. (1982) *Estudios Geol.* 38, 3.   {S 3.2}

George, M.T. and Vaidyan, V.K. (1981a) *J. Crystal Growth* 53, 300   {S 5.8}

George, M.T. and Vaidyan, V.K. (1981b) *J. Appl. Cryst.* 14, 345.   {S 5.8}

George, M.T. and Vaidyan, V.K. (1982a) *Cryst. Res. and Technol.* 17, 313.   {S 5.8}

George, M.T. and Vaidyan, V.K. (1982b) *J.Appl. Electrochem.* 12, 359.   {S 5.8}

Gerrard, J.E., Hoch, M., and Meeks, F.R. (1962) *Acta Metallurgica* 10, 751.   {S 1.1}

Ghez, R. (1988) *A Primer of Diffusion Problems.* Wiley-Interscience, New York.   {S 1.2}

Ghez, R. and Langlois, W.E. (1986) *Amer. J. Phys.* **54**, 646.   {S 1.3}
Glocker, D.A. and Soest, I.F. (1969) *J. Chem. Phys.* **51**, 3143.   {S 6.1}
Gould, H. and Tobochnik, J. (1988) *Computer Simulation Methods*, Addison-Wesley Publishing Co. Reading, Ma.   {S 1.1}
Halberstadt, E.S. (1967) *Nature* **216**, 574.   {S 6.1}
Happel, P., Liesegang, R.E., and Mastbaum, O. (1929a) *Kolloid Z.* **48**, 80.   {S 5.8}
Happel, P., Liesegang, R.E., and Mastbaum, O. (1929b) *Kolloid Z.* **48**, 252   {S 5.8}
Hatschek, E. (1911) *Kolloid Z.* **8**, 13.   {S 6.1}
Hatschek, E. (1914) *Kolloid Z.* **14**, 115.   {S 1.1}
Henisch H.K. and García-Ruiz, J.M. (1986) *J. Crystal Growth* **75**, 195.   {S 3.2}
Henisch, H.K. (1970) *Crystal Growth in Gels*. The Pennsylvania State University Press, University Park, Pa.   {S 1.1}
Henisch, H.K. (1988) *Crystals in Gels and Liesegang Rings*. Cambridge University Press.   {S 1.1, S 3.1}
Hermans, J.J. (1947) *J. Colloid Sci.* **2**, 387.   {S 6.1}
Holmes, H.N. (1926) *Colloid Chemistry*. (Editor: J. Alexander) Chemical Catalog Co. New York.   {S 6.1}
Joshi, M.S. and Antony, A.V. (1980) *Bull.Mat.Sci.* **2**, 31.   {S 6.1}
Kai, S., Müller, S.C., and Ross, J. (1982) *J.Chem.Phys.* **76**, 1392.   {S 6.2}
Kai, S. Müller, S.C., and Ross, J. (1983) *J.Chem.Phys.* **87**, 806.   {S 6.2}
Kanniah, N. (1983) *Revert and Direct Liesegang Phenomenon*. Thesis, Crystal Growth Center, Anna University, Madras, India.   {S 4.3}
Kirov, G.K. (1969) *Compt. Rend. Acad. Bulg. Sci.* **22**, 915.   {S 3.2}
Kirov, G.K. (1972) *J. Crystal Growth* **15**, 102.   {S 3.2, S 6.1}
Kisch, B. (1929) *Kolloid Z.* **49**, 154 and 156.   {S 5.8}
Knöll, H. (1938a) *Kolloid Z.* **82**, 76.   {S 4.2}
Knöll, H. (1938b) *Kolloid Z.* **85**, 290.   {S 4.2}
Knöll, H. (1939) *Kolloid Z.* **89**, 135.   {S 4.2}
Kratochvil, P., Sprusil, B., and Heyrovsky, M. (1986) *J. Crystal Growth* **3/4**, 360.   {S 6.1}
Lendvay, E. (1964) *Acta Phys.Hung.* **17**, 315.   {S 3.2}
Liesegang, R.E. (1896) *Naturwiss. Wochenschrift* **11**, 353.   {S 1.1}
Liesegang, R.E, (1897) *Z. phys. Chem.* **23**, 365.   {S 1.1}.
Liesegang, R.E. (1898) *Chemische Reaktionen in Gallerten*, Düsseldorf.   {S 1.1}
Lovett, R., Ross, J., and Ortoleva, P. (1978) *J.Chem.Phys.* **69**, 947.   {S 6.2}

Milne, W.E. (1953) *Numerical Solution of Differential Equations*, John Wiley and Sons, New York. Also (1970) Dover Publications Inc. {S 1.4}

Mullin, J.W. (1961) *Crystallization*, Butterworths Scientific Publications, London.   {S 3.1}

Matalon, R. and Packter, A. (1955) *J. Colloid Science* **10**, 46.   {S 4.3}

Mathur, P.B. (1961) *Bull.Chem.Soc. Japan* **34**, 437.   {S 4.3}

Morse, H.W. and Pierce, G.W. (1913) *Z. phys.Chem.* **45**, 589.   {S 4.3}

Miyamoto, S. (1937) *Kolloid Z.* **78**, 23.   {S 5.8}

Nickl, J. and Henisch, H.K. (1969) *J. Electrochem. Soc.* **116**, 1258. {S 6.1}

Ostwald, Wi. (1897a) *Z. phys. Chem.* **22**, 289.   {S 3.1}p

Ostwald, Wi. (1897b) *Z. phys. Chem.* **27**, 365.   {S 3.1}

Prigogine, I. (1984) *Order out of Chaos; Man's Dialogue with Nature.* Bantam Books, New York.   {S 1.1}

Packter, A. (1055) *Nature* **175**, 556.   {S 4.3}

Rainville, E.D. (1963) *The Laplace Transform.* The Macmillan Co. New York.   {S 1.2}

Ross, J. (1984) *Physica* **12D**, 303.   {S 6.2}

Shewmon, P.G. (1963) *Diffusion in Solids.* McGraw-Hill Book Co., New York.   {S 1.2}

Salvinien, J. and Moreau, J.J (1960) *J. Chim. Phys.* **57**, 518.   {S 3.2}

Shinohara, S. (1970) *J. Phys. Soc. Japan.* **29**, 1073.   {S 4.3}

Shinohara, S. (1974) *J. Phys. Soc. Japan.* **37**, 264.   {S 4.3}

Van Hook, A. (1938) *J. Phys. Chem.* **42**, 1191.   {S 3.2}

Van Hook, A. (1963) *Crystallization; Theory and Practice.* Reinhold, New York.   {S 3.1}

Wagner, C. (1950) *J. Colloid Sci.* **5**, 85.   {S 4.3}

Zettlemoyer, A.C. (1969) *Nucleation.* Dekker, New York.   {S 3.1}

ORDER FORM.

PERIODIC PRECIPITATION SOFTWARE

Programs PPBIN**, PPMON** and PPCPG**, as described in Appendices A, B, C, which serve as documentation. Hardware requirements: IBM-compatible microcomputer, with EGA or VGA color display.

Available in a "standard version", ready-to-run without any other software. Also available with the True BASIC™ source code in ASCII files, which can be modified by the user and adapted to special needs. This requires the True BASIC Language System, with its built-in editor and compiler.

Standard version:_____(5¼" floppy)................$65.00
                 _____(3½" floppy)................$65.00

Source code version:_____(5¼" floppy).............$100.00
                    _____(3½" floppy).............$100.00

True BASIC™ Language System, ver. 2.1_____$99.95

Postage and packing...$ 4.00
(1991-92 prices) Institutional orders or prepaid US$ checks. Pa. residents please add 6% sales tax.

NAME_____TEL_____

STREET_____

CITY_____STATE____ZIP_____

COUNTRY_____

The Carnation Press
P.O.Box 101,
State College, Pa. 16804

INDEX

accumulation, mass 50 et seq.
accuracy, computational 12
agar 3
alcohol 84
algorithm 4, 11 et seq. 17 et seq.
alumina 2
ammonium phosphate 83
analytic solutions 17
anisotropy 2
attempt frequency 7, 8
averaging 50 et seq., 107

bacterial deposits 42
banding 86, 96
Boltzmann's constant 30

calcium 2, 4
ceramics 2
ceratostigma plumbaginoides 2
chaos 1, 90
charge density 10
chromates 2, 3, 4, 11
closure, reservoir 53 et seq., 62, 104
competitive particle growth 86 et seq., 90 et seq.
concentration decrements 33 et seq., 38, 113
concentration increments 38, 113
concentration product 25 et seq., 43, 58, 60 et seq. 74
contamination, reservoir 53 et seq.
copper 2
critical (size) nucleus 30
crystal growth 1
cylindrical coordinates 10

density, particle 112 et seq.
depletion, reservoir 53 et seq.
deposits, late 68
deterministic nucleation 46 et seq., 105
diffusion column 53, 61
diffusion constant 8, 12 et seq., 19 et seq., 67 et seq.

diffusion constant, conc. dependence 9 et seq., 14, 23
diffusion constant, unequal 67, 68
diffusion currents 10
diffusion length 12
diffusion theory 7 et seq.
diffusion, non-homogeneous media 9 et seq., 20 et seq.
diffusivity; see diffusion constant
dimensions 12, 35
display 6, 58, 59, 104, 107, 108
display, numerical 108

Einstein relationship 13
electric fields, alternating 77
electric fields, static 9 et seq., 24 et seq., 76 et seq., 106
electric fields, temporary 78, 106
energy, free 30
epidermis, leaf 2, 5
equality range 32 et seq., 66, 73 et seq.
equilibrium concentration 88 et seq., 112
experimentation, hypothetical 1, 7, 57 et seq., 83 et seq.

Fick's laws of diffusion 7 et seq.
field application time 106
field currents 10
field removal time 106
fields, electric; see electric fields
flux 7, 8
frequency, attempt 7, 8

gel 2
gold 83, 88
grain count 33 et seq., 59 et seq.
grain size 33 et seq., 52, 59 et seq.
graphics 103
growth 30, 37 et seq., 64, 86 et seq.
growth coefficient 37 et seq., 89, 99, 112

half-period 77, 106

hydrogen potential (pH) 35, 36, 51 et seq.

iodides 1, 3, 88

lead 1, 3
lead iodide, 88
Liesegang R.E. 1, 2, 4, 41, 47

mass per grain 33 et seq., 38, 105, 109, 111 et seq.
mass profile 47, 59, 62, 66 et seq., 75, 85, 90 et seq., 108
membranes 8
methods A and B 48, 80
mobility, electrical 10 et seq.
monomer systems 2 83 et seq. 110 et seq.
moving deposits 44, 85, 87
multiple runs 49 et seq., 105

non-homogeneity 9, 20 et seq.
normalization 13, 35
nucleation 29 et seq.
nucleation coefficient, deterministic 48 et seq.
nucleation cutoff 33
nucleation, heterogeneous 30
nucleation probability 32, 46, 79 et seq., 105

Ostwald ripening 86 et seq.

particle density 112 et seq.
particle radius 88 et seq., 111 et seq.
phase reversal 106
phosphates 2, 4
Poisson's equation 10
polysaccharides 5
potassium dihydrogen phosphate 83
pre-charging 67, 104
precipitates, periodic (examples) 1 et seq.
precipitation product 29 et seq., 57, 60, 105
precipitation product, variable 68, 70
printout, hardcopy 105, 108, 109
probabilistic mode 46 et seq., 79 et seq.
profiles, concentration 17 et seq.

pseudo-language 6

quasi-stability 72

radial symmetry 10, 14
random walk 7, 12, 45 et seq., 85, 86
re-solution 6, 35, 37 et seq., 64, 85
re-solution coefficient 37, 39, 40
reactions, oscillatory 1
reservoir closure 53 et seq.,62, 104
reservoir depletion coefficient 54, 55, 60 et seq.
reservoirs 15, 17, 53 et seq., 60, 62, 90, 111
revert deposit patterns 46

saturation 29 et seq., 84 et seq., 109
secondary products, effect of 52, 70 et seq., 105
secondary reactions 51 et seq., 60, 70 et seq.
semiconductors 8
sensitivity factor 84, 109
silica 2, 61
silver 2 et seq.
silver chromate 4, 11
silver nitrate 5
simulation methods 5, 81
software 6, 7, 57 et seq., 103 et seq.
sols 88
solubility modulation 83 et seq., 87
solubility product 29, 35, 59, 60, 105
solubility product, variable 68 et seq.
spacing relationships 44 et seq., 63
stoichiometry 31 et seq., 66, 73 et seq.
subroutines 103 et seq.
supersaturation 29 et seq., 84 et seq., 109

True BASIC (language) 6, 16, 103

urea 42
user interface 58 et seq., 103 et seq.

video update 105

MAR 1 5 1991